Contents

Working Party members

Mr G. Lionel Wright (Chairman)
Mr Arthur Appleton
Mr Peter Baldwin
Mr Gordon Bateman
Mr Roger Button
Mr R. Brian Carter
Mr John Challenger
Mr Mike Foster
Mr Geoffrey Hawker
Mr Robert Kimber
Mr Colin Pim
Mr Henry Rowson
Mr David Wright

The Working Party dedicates this edition to Peter Baldwin, who sadly passed away just before work on the book was completed. Peter had been a much-valued member of the Working Party since its early days and his tireless dedication and immense contribution continued over many years until shortly before his death. He will be missed greatly.

Introductory notes

1. Use of this handbook

This handbook (which is often referred to as the Institution of Chemical Engineers' (IChemE) 'Green Book') contains a form of Agreement and a set of General Conditions of Contract for reimbursable contracts. Guidance is provided on how to compile the Specification and the Schedules to which reference is made in the General Conditions. Guide Notes A to T are intended to aid interpretation of some of the General Conditions and to explain the possible need for certain Special Conditions. Users of this Form of Contract are encouraged to read the relevant guidance sections before preparing the Specification, the Schedules and any Special Conditions, all of which will have to be written individually. The Specification and most of the Schedules form essential parts of a workable contract, although a few of the Schedules may be considered to be optional.

2. General

This handbook has been developed by a working party appointed by IChemE, the members of which represent a broad spectrum of those involved in the specification, design, construction, ownership and operation of process plants. The General Conditions in particular have been formulated to reflect the practices and relationships pertaining to the process plant sector, which are generally far less adversarial in nature than in other parts of the construction industry. Anyone contemplating modifications to the General Conditions, or the addition of Special Conditions to modify or amplify them, should take care not to introduce provisions which may conflict with these well-established practices and relationships. They should particularly be aware of the risk of introducing inconsistencies within the Conditions, or provisions which may be unenforceable.

Users of these General Conditions are cautioned that IChemE has not authorised or approved any amendments to any of the provisions unless they are published by IChemE.

These General Conditions have been drawn up with process plants in mind. While they have been used successfully for projects outside the process industries, great care should be taken to draft provisions which properly reflect differences in custom and practice for any part of the works for which the preparation of non-performance related engineering work differs significantly from that normally found in a typical process plant project.

These General Conditions incorporate at Clause 39 provisions regarding payment which are consistent with the requirements of the Housing Grants, Construction and Regeneration Act 1996, Part II, and Clause 46 provides for adjudication in accordance with the Act.

The Form of Contract as drafted is for a project carried out within the United Kingdom and subject to English law. However, it has become increasingly popular for use in connection with projects outside the United Kingdom. Guidance on the use of this Form of Contract for these projects is provided in Guide Note U.

3. Choice of contract–lump sum or reimbursable?

Contracts based on 'lump sum' and wholly 'cost reimbursable' payment are the extremes of a wide range of forms of contract that can be used to cover the distribution of risks and responsibilities between the parties to an engineering and construction contract for the realisation of a process plant. Whilst, at first sight, it would appear that the Purchaser's best interests would be served by the use of a contract based upon a lump sum price for his project, he must be quite sure that he has properly examined the alternatives and assessed their relative advantages and disadvantages.

The essence of a lump sum contract is that, in return for an agreed firm price, the Contractor provides all that is necessary to ensure that the finished plant complies with the specification, achieves the required levels of performance and is completed on time. The specification, performance requirements and guarantees, and all other technical and commercial contractual conditions, must be agreed in detail before the contract is awarded. Once the contract becomes effective, the lump sum price is payable regardless of the actual costs incurred by the Contractor. The Contractor must, therefore, include contingencies in his price for estimating errors, uncertainties such as cost escalation and currency fluctuations, costs which, on the basis of his experience and despite all efforts to avoid them, will inevitably be incurred from time to time such as those arising from design, procurement or construction errors, and any other unforeseen costs that he might incur, but has no protection against these risks under the terms of the contract. Having included for these contingencies, the eventual cost of which may turn out to be significantly different from his estimates, the Contractor takes the risk that over-expenditure will result in a smaller profit than expected or even a loss, in the knowledge that, if all goes well, his profit may be greater then expected. Conversely, whilst the Purchaser does not carry the risk of cost overruns, he cannot benefit from any savings.

In contrast, the essence of a reimbursable contract is that in providing the goods and services that the Purchaser requires, the Contractor is reimbursed for all the costs he incurs including, subject to the contract and provided he exercises normal professional skill, corrective costs arising from the inevitable design and construction errors. The Purchaser gains the benefits of any engineering, material or construction cost savings but bears the risk of any cost over-run.

Table 1 compares the advantages and disadvantages of lump sum and reimbursable contracts and provides guidance upon which the choice of contract form may be based. If the Purchaser elects to place a lump sum contract it is essential that he provides the potential contractors with adequate information upon which to base their estimates and adequate time to submit their tenders.

Although initial lack of definition of the Purchaser's requirements may well be a reason why a particular contract is let on a reimbursable rather than a lump sum basis, it may not be the only factor that influences the decision. In fact there is no reason why a project which is completely defined should not be let on a reimbursable basis.

Table 1—Comparison of lump sum and reimbursable contracts

LUMP SUM	REIMBURSABLE
Project definition required in enquiry	
• Complete.	• Minimum—sufficient to enable tenderers to check that they have the resources needed.
Advantages	
• Purchaser knows his expenditure commitment. • Clarity of contractual risk and project management. • Allows fullest competition between potential contractors. • Bid evaluation is more straightforward.	• Requires minimum enquiry definition. • Shortest possible bid time. • Complete flexibility—design development and Purchaser participation practicable. • Purchaser/Contractor conflict of interest is minimised. • Purchaser has control over costs incurred. • Purchaser can assess tenderers' rates. • Purchaser can use Contractor to evaluate alternative schemes. • Purchaser can terminate at will without incurring substantial costs. • Competition can continue for subcontract packages, when developed.
Disadvantages	
• Purchaser/Contractor interests are more divergent than in other forms of contract. • Lengthy enquiry preparation time to prepare clear and complete specification and bid documentation. • Long bidding time is required (2–4 months or more). • Crucial process design phase compressed into a very short period. • Lack of flexibility—changes are difficult/expensive. • Purchaser participation in project is difficult. • Cost to Purchaser may be unnecessarily high due to contingencies for risk and escalation. • Emphasis on low bid price in bid evaluation may divert attention from technical merit.	• Contractor has no monetary incentive to minimise cost to Purchaser. • Purchaser has no assurance of final cost. • Purchaser has to check and verify Contractor's manhour and expense records. • Bid evaluation may be difficult.
Typical applications	
• When competing tenderers offer different proprietary processes. • When Purchaser provides tenderers with front-end engineering packages and detailed specifications. • 'Standard package' plants.	• Where flexibility is required, such as when project cannot be defined in detail at the start because not yet fixed. • When Purchaser wishes to participate extensively in the design—for example, a development project, especially one dependent on Purchaser know-how. • When confidentiality considerations preclude the issue of detailed enquiry specifications to tenderers. • When a very short programme is required. • When Purchaser considers contingencies for risk in a lump sum are likely to be excessive. • When potential contractors are not prepared to take lump sum contracts (for example, very large contracts, periods of high escalation, times of high inflation).
Comments	
• Complete project definition is essential for a lump sum contract. • Tenderers' tendering costs are very high. Purchaser should minimise contracting industry tendering costs (and thus plant costs) by (a) not inviting bids until there is a high probability of the project proceeding, (b) minimising the number of tenderers, (c) pre-qualifying each tenderer so that he will not be rejected later on grounds that should have been known to Purchaser before the enquiry was issued, and (d) considering reimbursing pre-qualified unsuccessful tenderers for their costs of bid preparation.	• The most flexible type of contract, allowing a very rapid start. • Flexibility may encourage Purchaser to introduce design changes, resulting in increased cost and longer programme. • Contractor's profit/loss is limited. • Conversion of all or part of contract to a lump sum basis is possible during the project at the stage when scope of work becomes fully defined.

In drafting its Forms of Contract, IChemE has adopted the basic philosophy that the parties should co-operate to achieve the mutual objective of a successful project rather than regarding the contract as the basis for an adversarial relationship. IChemE believes that it is in the best interests of the parties to deal fairly with each other and with the contractors, subcontractors, specialists and suppliers in an atmosphere of co-operation in order to achieve 'win-win' solutions to problems that will inevitably arise during the course of the project.

There is a variety of other forms of payment between the two extremes which may form the basis of the contract. Typically, these range as follows:

- reimbursable cost plus a percentage fee;
- reimbursable cost plus a fixed fee;
- target cost (shared over-run and/or under-run);
- unit rate (including remeasure);
- guaranteed maximum price;
- lump sum services and materials with reimbursable construction;
- lump sum (i.e. wholly lump sum).

Only the first of these is wholly reimbursable but all except the last involve, to some extent at least, reimbursement to the Contractor of the costs he actually incurs.

In any partly reimbursable contract, the provision of some goods and/or services is covered by a lump sum payment while others are paid for on a reimbursable basis. The most commonly used combinations are given in Table 2.

Table 2—Classes of partly reimbursable contracts

Major cost element	1	2	3	4	5	6	7
A—Home office engineering	✓	✓	✓	×	×	×	×
B—Home office construction	✓	✓	✓	×	×	×	×
C—Home office procurement	✓	✓	✓	✓	×	×	×
D—Materials	✓	✓	✓	✓	✓	✓	×
E—Site office	✓	✓	✓	✓	✓	×	×
F—Site labour	✓	✓	✓	✓	✓	✓	✓
G—Site equipment	✓	✓	✓	✓	✓	✓	✓
H—Overheads	✓	✓	×	×	×	×	×
J—Profit (and licence fee, if applicable)	✓	×	×	×	×	×	×

Notes

(a) ✓ means the Contractor's costs for this cost element are reimbursed. In Class 1 (wholly reimbursable) the overhead and profit elements (items H and J) would be agreed percentages of base sums comprising one or more of the items A–G, while any licence fee would be reimbursable at net cost.

(b) × means the Contractor includes his price for this cost element within his lump sum quotation.

(c) Classes 2–7 are all reimbursable contracts in which the Contractor provides a total service, but one or more of the elements is covered by a lump sum.

Suggestions for the preparation of contract documentation for 'mixed' contracts, where parts of the payments to the Contractor are made as lump sums may be found in the Guidance on compiling Schedule 18 (Cost elements, rates and charges), under the heading 'Lump sum items'. There is a distinction to be drawn between work carried out by the Contractor himself and by his Subcontractors, in that the payment of the latter by lump sums can be achieved by using a form of subcontract such as the Yellow or Brown books published by IChemE, which provide, for example, for the Subcontractor to make good defects at his own expense, and also to pay damages for performance failure or delay in completion.

It will be clear that for any form of reimbursable contract the cost elements need to be clearly defined together with the means by which payments due to the Contractor are to be calculated. This is the purpose of Schedule 18 (Cost elements, rates and charges), which has to be used with these General Conditions.

While attracted to the flexibility of a cost reimbursable contract, some Purchasers may prefer the creative tension and discipline which can be engendered by the introduction of a target cost into the Contract. In deciding to adopt a target cost contract, the Purchaser must consider the point within the programme of work at which the plant definition is sufficient for the Contractor to give a firm target cost.

Adopting the Green Book Form of Contract to accommodate a target cost requires careful alteration to some of the clauses and to Schedule 1 (Description of the Works) and 19 (Terms of Payment) to avoid conflict within the contract. A supplement to the Green Book covering Target Cost Contracts is published by IChemE giving guidance on drafting such a contract.

4. Conditions of Contract

The General Conditions of Contract (the 'Conditions') in this handbook have been written for reimbursable contracts. Purchasers who wish to consider the use of any form of lump sum contract should refer to the IChemE Form of Contract for Lump Sum Contracts (the 'Red Book'), which is based on the same philosophy as and published in parallel with this handbook.

The Conditions have been drawn up primarily for process plants, i.e. those dealing with chemical and/or physical changes of materials in bulk. Such plants usually consist of a number of process equipment items with intermediate storage facilities, inter-linking pipework or conveyors, structures, instrumentation and control systems, and auxiliary mechanical and electrical equipment and safety or environmental protection equipment. There will also be foundations, buildings and other civil engineering works, utility services, storage facilities, and off-sites although these may be part of another contract. A process plant is further characterised by the need to test for and demonstrate its ability to operate as an integrated unit to manufacture the products for which it is designed and when supplied with specified feedstocks, raw materials and utilities to meet parameters such as the quantity and quality of products, by-products and effluents produced. Product yields and efficiencies or their equivalent, utility consumptions, or any other appropriate chemical engineering criteria of performance are also normally the subject of tests and appropriate guarantees, in addition to mechanical soundness under operating conditions.

The carrying out of a process plant project involves some or all of the following activities: comparative evaluations of different processes; chemical engineering design; detailed engineering in all disciplines; supply of the hardware; site construction work; start-up and plant operation. However, the extent to which these are within the scope of particular contracts, the nature of different process plants and their special circumstances including process rights varies widely. Thus, each contract may be considered to be unique and to require a tailor-made set of contract documents within the general framework of provisions appropriate to virtually all process plants.

5. Precontract activities

Comparison of tenders for reimbursable contracts is not easy but is facilitated by the Purchaser ensuring that each covers the same scope of work. Frequently because of the difficulty of balancing the different amounts and rates the final decision will depend on the Purchaser's assessment of the technical and management capability of the Contractors. It is therefore important to ensure that past records of the Contractors in the relevant field of technology are known.

The documents comprising this form of contract contain a wealth of detail and will need considerable resources of experienced project personnel and time to compile properly.

The tender invitation should therefore incorporate a draft contract document including as a minimum drafts of the schedules containing the Purchaser's information and requirements:

- Specification;
- description of the Works;
- responsibilities of Purchaser;
- health and safety;
- quality assurance and validation (if applicable);
- subcontracting;
- training by Contractor;
- times of completion;
- liquidated damages for delay;
- performance guarantees and damages for failure;
- terms of payment;
- cost elements, rates and charges.

Outline frameworks for the other schedules should also be included to be completed by the tenderers.

The better the definition in the tender invitation, the shorter will be the time needed for clarification and evaluation of the tenders. For example, if the Purchaser wishes to benefit from the application of Schedule 10 (Parts with limited working life) this should be made clear in the tender invitation.

A similar consideration applies to factors which will affect the risks of the contractor who will eventually be appointed. These include the amounts to be included in paragraphs 3, 4 and 5 of the Agreement, as should any proposal by the Purchaser to make incentive payments (see Guide Note H) or to introduce a target cost (see the Green Book supplement on Target Cost Contracts).

If the Purchaser would prefer to carry the risk of inflation in respect of any price to be included in the Contract, the method of cost adjustment he proposes should be indicated in the invitation to tender in broad terms only, the tenderer being invited to submit his own proposals.

If the tenderer cannot accept the Purchaser's requirements in full or if there is any doubt about the compatibility of his own proposals and those of the enquiry, he should bring this to the Purchaser's attention at or before the time of tendering. The points at issue should be discussed and a course of action agreed. Tender evaluation is assisted if in such cases the tenderer finds it possible to submit alternatives incorporating his own proposals in addition to a tender based exactly on the terms of the enquiry.

6. Contract procedure

It is strongly recommended that the formal Agreement be used, in the form provided. This includes a number of statements which are important for clear definition of the terms of the Contract. Quite apart from stating a limit to the Contractor's liability under Sub-clause 30.7, the amount of the deductible payable under the insurance arranged by the Purchaser, and naming the Project Manager and the Contract Manager, the Agreement defines exactly which documents make up the Contract. Documents dated earlier than the date of the Agreement would normally have no effect unless specifically incorporated in the Contract nor would any other terms and conditions, for example those sometimes printed on the back of order forms or quotations.

If any other form of agreement is to be used the Purchaser must ensure that all the statements shown in the Agreement in this form of Contract are included in the contract documentation in some other way. It is essential that all the necessary supporting documents, including the General and Special Conditions of Contract, with the accompanying Schedules, and any other necessary contractual and technical documents are incorporated and that a comprehensive contents list is provided, similar in form to the Annex to the Agreement. There are frequent references to the Agreement within the General Conditions.

When there have been any changes in concept during precontract negotiations, any relevant Special Conditions or Schedules should be amended before the Agreement is signed.

7. Modifications to the Contract

Any part of the contract documentation can be modified by mutual agreement between the parties at any time after it has been signed, but any modification must be expressed in unambiguous terms in a document signed by both parties.

For example, if a part of the contract documentation has not been finalised at a time when it is in the interests of both parties for the Agreement to be signed and for the Contractor to start work, it is still possible to complete the missing details by agreement. A prime example would be a contract where the Specification merely contained the performance criteria for the Plant and did not list the component parts of the Plant, because it was intended that the Contractor should develop process details in the course of carrying out his design work. In that situation it would only be after the equipment for the Plant had been properly defined that the details of any appropriate off-site tests could be included in Schedule 13 (Pre-installation tests and procedures).

Agreement

(See Guidance on Completing the Agreement)

Contract Title and Identification Number:

THIS AGREEMENT is made the *[day, month and year in words]*

between

of

(hereinafter called 'the **Purchaser**') of the one part

and

of

(hereinafter called 'the **Contractor**') of the other part.

WHEREAS

The **Purchaser** wishes to have a process plant to be known as

constructed at

and wishes the **Contractor** to carry out and complete the **Works** as defined in the **Contract** and the **Contractor** is willing and able to carry out and complete the **Works** in accordance with the **Contract**.

THIS AGREEMENT provides as follows:

1. The following documents and their attachments (if any), shall together constitute the contract between the **Purchaser** and the **Contractor** and the term '**Contract**' shall in all such documents be construed accordingly.

   (a) This Agreement.
   (b) The Special Conditions (if any).
   (c) The General Conditions of Contract being Clauses 1–48 as set out in the IChemE Form of Contract for Reimbursable Contracts (the 'Green Book'), 3rd edition, 2002.
   (d) The Specification.
   (e) The Schedules:
   Schedule 1: Description of the Works;
   Schedule 2: Documentation;
   Schedule 3: Responsibilities of Purchaser;
   Schedule 4: Health and Safety;
   Schedule 5: Environmental protection and waste disposal;
   Schedule 6: Quality assurance and validation;
   Schedule 7: Subcontracting;
   Schedule 8: Contractor's named personnel;
   Schedule 9: Training by Contractor;
   Schedule 10: Parts with limited working life;
   Schedule 11: Times of completion;
   Schedule 12: Liquidated damages for delay;
   Schedule 13: Pre-installation tests and procedures;
   Schedule 14: Criteria for the completion of construction;
   Schedule 15: Take-over procedures;
   Schedule 16: Performance tests and procedures;
   Schedule 17: Performance guarantees and damages for failure;
   Schedule 18: Cost elements, rates and charges;
   Schedule 19: Terms of payment;
   Schedule 20: Commercial records.

   For the purpose of identification, the contents of the **Contract**, including the number of pages in each part, are listed in the Annex to this Agreement.

2. The **Contract** constitutes the entire agreement between the **Purchaser** and the **Contractor** with respect to the performance of the **Works** and supersedes all prior negotiations, representations or agreements relating thereto, whether written or oral, except to the extent that they are expressly incorporated in the **Contract**. No changes, alterations or modifications to the **Contract** shall be effective unless the same shall be in writing and signed by both parties.

3. The **Contractor's** liability in respect of loss of or damage to property of the **Purchaser** and his **Affiliates** in accordance with Sub-clause 30.7 of the General Conditions shall not exceed £ *[amount in figures and words].*

4. The deductible for the insurance policy referred to in Sub-clause 31.2 of the General Conditions shall be the first £ *[amount in figures and words]* of each and every claim.

5. The total aggregate liability of the **Contractor** to the **Purchaser** under Sub-clause 44.3 shall not exceed £ *[amount in figures and words].*

6. In case of conflict between any of the documents accompanying this Agreement, the order of precedence shall be as set forth in Clause 2 of the General Conditions.

7. For the purposes of Sub-clauses 7.3, 8.2 and 28.3 of the General Conditions the date of the **Contractor's** tender shall be *[date].*

8. The date for the commencement of the **Works** shall be *[date].*

9. The **Purchaser** hereby appoints *[name]* to act as the **Project Manager** for the purposes of the **Contract**.

10. The **Contractor** hereby appoints *[name]* to act as the **Contract Manager** for the purposes of the **Contract**.

11. The bank whose base lending rate is referred to in Clause 1 of the General Conditions is *[name of bank].*

12. The location of any arbitration proceedings under Clause 48 shall be *[place].*

IN WITNESS whereof the parties hereto have signed this Agreement on the date first above written.

For and on behalf of *[full name of the **Purchaser**]*

Signature:

Name:

Position:

For and on behalf of *[full name of the **Contractor**]*

Signature:

Name:

Position:

General Conditions of Contract

1. Definition of terms

In the **Contract**, unless the context otherwise requires, the following expressions shall have the meanings hereby assigned to them.

'**Acceptance Certificate**' has the meaning set out in Sub-clause 36.3 and '**Acceptance**' and '**Accepted**' shall be construed accordingly.

'**Affiliate**' means any company which is either directly or indirectly controlled by a party hereto or its ultimate parent, where control is signified either by having effective control of the appointment of a majority of a company's board of directors or by the beneficial ownership of more than half of the nominal value of its equity share capital.

'**Agreed Rate**' means the base lending rate as may be current from time to time of the Bank stated in the Agreement or such rate as agreed by the parties for the currency applicable to the **Contract**.

'**Applications Software**' means software that has been developed specifically as part of the **Works** including configuration of **Standard Software** or **Contractor's Software.**

'**Approved Programme**' means the programme of work approved in accordance with Clause 13 (Times of completion and Approved Programme).

'**Confidential Information**' has the meaning set out in Sub-clause 20.1.

'**Contract**' has the meaning set out in the Agreement.

'**Contract Manager**' means the individual named as such in the Agreement subject to Clause 12 (Contract Manager and Contractor's staff).

'**Contract Price**' means the total of the sums payable by the **Purchaser** to the **Contractor** in accordance with Clause 18 (Contract Price).

'**Contractor**' means the person named as such in the Agreement or his permitted assigns.

'**Contractor's Equipment**' means all equipment, construction plant, vehicles, temporary facilities, materials, tools or things brought on to the **Site** by or on behalf of the **Contractor** for carrying out the **Works** but not for permanent incorporation in the **Plant**.

'**Contractor's Software**' means software developed by the **Contractor** other than **Applications Software**.

'**Decision**' means any decision, certificate, notice, instruction, order, agreement, approval, rejection or consent.

'**Defect**' has the meaning set out in Sub-clause 37.2.

'**Defects Liability Period**' has the meaning set out in Sub-clause 37.2.

'**Documentation**' means any relevant documents in paper or electronic form, including drawings, technical software, images, designs or records.

'**Expert**' means a person referred to in Clause 47 (Reference to an Expert).

'**Final Certificate**' has the meaning set out in Sub-clause 38.1.

'**IChemE**' means the Institution of Chemical Engineers.

‘**Legislation**’ means all applicable laws, statutes, bye-laws, regulations and other measures having the force of law.

‘**Materials**’ means machinery, plant and other items of equipment and materials intended to form part of the **Plant** and other things needed for its operation, to be supplied by the **Contractor**.

‘**Plant**’ means the plant as described in the **Specification** to be constructed at the **Site**.

‘**Project Manager**’ means the individual named as such in the Agreement subject to Clause 11 (The Project Manager).

‘**Project Manager’s Representative**’ has the meaning set out in Sub-clause 11.6.

‘**Purchaser**’ means the person named as such in the Agreement or his permitted assigns.

‘**Site**’ means the area within which the **Plant** is to be constructed, together with all other areas as the **Contractor** shall be permitted to use in connection with the **Works**, as specified in Schedule 1 (Description of the Works).

‘**Site Manager**’ has the meaning set out in Sub-clause 12.2.

‘**Specification**’ means the Specification issued as part of the **Contract**, which sets out the technical definition of the **Plant**.

‘**Standard Software**’ means software which is proprietary to a third party.

‘**Subcontractor**’ means any subcontractor of any tier to whom the preparation of any design, the supply of any **Materials** or the carrying out of any other part of the **Works** is subcontracted.

‘**Suspension Order**’ has the meaning set out in Sub-clause 41.1.

‘**Take-Over Certificate**’ has the meaning set out in Sub-clause 33.7.

‘**Variation**’ has the meaning set out in Sub-clause 16.1.

‘**Variation Order**’ means an order by which a **Variation** is ordered or other notification made to the **Contractor** in accordance with the **Contract**.

‘**Works**’ means the design, engineering and other services to be provided by the **Contractor** including, but not limited to, the provision and construction of the **Plant** and any temporary works, and any other work to be carried out by the **Contractor** in accordance with the **Contract**.

2. Interpretation

2.1 Unless otherwise agreed, the **Contract** shall be governed and construed according to the laws of England.

2.2 The **Contract** documents shall be construed as mutually explanatory of one another. In the event of conflict between any of the documents comprising the **Contract**, the order of precedence shall be as follows:

(a) the Agreement;

(b) the Special Conditions;

(c) these General Conditions of Contract;

(d) the **Specification**;

(e) the Schedules.

2.3 The singular shall include the plural and the plural the singular except where the context otherwise requires and the words ‘he’, ‘him’ and ‘his’ shall be taken to mean ‘she’, ‘her’ and ‘hers’ where appropriate.

2.4 No approval or consent required to be obtained under the **Contract** shall be unreasonably refused or delayed.

2.5 'Day' shall mean a calendar day.

2.6 Any communication which could have a contractual effect shall be in writing, which shall include handwritten, typewritten or printed documents sent by hand, post, fax, secure e-mail or other means resulting in a permanent record.

2.7 The headings and marginal notes in these General Conditions of Contract and in any Special Conditions, together with the accompanying Introductory Notes, Guidance on compiling the Agreement, **Specification** and Schedules and the Guide Notes, shall not form part of the **Contract** nor shall they be taken into consideration in the interpretation or construction thereof.

3. Contractor's responsibilities

See Guidance on compiling Schedules 1 (Description of the Works) and 6 (Quality assurance and validation)

3.1 In consideration of payment by the **Purchaser**, the **Contractor** shall carry out and complete the **Works**, all in accordance with the provisions of the **Contract**.

3.2 Subject to the express provisions of the **Contract**, all work carried out by the **Contractor** under the **Contract** shall be carried out with sound workmanship and materials, safely and in accordance with good engineering practice, applicable **Legislation** and codes and shall be to the reasonable satisfaction of the **Project Manager**.

3.3 The **Contractor** shall set out the **Plant** by reference to points, lines and levels of reference provided to him by the **Project Manager**.

3.4 The **Plant** as completed by the **Contractor** shall comply with the **Specification** and any other requirements of the **Contract**.

3.5 If at any time during the performance of the **Contract** the **Contractor** is of the opinion that a change to the **Works**, or the design or method of operation of the **Plant**:

(a) is necessary to eliminate a potential defect in the **Plant** or a specific hazard to any person or party in the performance of the **Works** or in the operation of the **Plant** which has occurred or would otherwise occur; or

(b) would otherwise be beneficial to the **Purchaser**

the **Contractor** shall bring the matter to the attention of the **Project Manager** stating the reasons for his opinion and where appropriate submit his proposals for a **Variation** in accordance with Clause 17 (Contractor's Variations).

3.6 The **Contractor** shall at all material times have and maintain resources (including financial resources) adequate to carry out the **Works**.

3.7 The **Contractor** shall provide and maintain a system of cost control, operated by suitably qualified and experienced personnel, to the satisfaction of the **Project Manager**. Such control system shall include:

(a) records of commitments of expenditure on orders placed;

(b) records of invoices paid;

(c) records of cost changes resulting from **Variation Orders**; and

(d) a forecast of the cost of the **Works** to their final completion, broken down into defined sections.

Unless otherwise agreed, the **Contractor** shall at intervals of not more than one calendar month report in writing to the **Project Manager** on the progress of the **Works**, supporting his reports with appropriate **Documentation** including cost control reports.

3.8 The **Contractor** shall maintain, and cause **Subcontractors** to maintain, a quality assurance system as described in Schedule 6 (Quality assurance and validation). The existence of such a quality assurance system shall not relieve the **Contractor** from any of his duties, obligations or liabilities under the **Contract**. The **Contractor** shall also prepare and implement a validation plan, if such a requirement is included in Schedule 6.

3.9 The **Contractor** shall provide training of **Purchaser's** personnel as described in Schedule 9 (Training by Contractor).

3.10 The **Contractor** acknowledges and accepts that, to the extent that he is the 'Designer' of the **Plant** under the CDM Regulations referred to in Clause 26.1, he is responsible for ensuring that the design shall be executed in full compliance with the said regulations.

4. Purchaser's responsibilities

See Guidance on compiling Schedule 3 (Responsibilities of Purchaser)

4.1 The **Purchaser**, through the **Project Manager**, shall provide the **Contractor** with the **Documentation** described in Schedule 3 and with all such further information and **Decisions** as are necessary to be provided by the **Purchaser** in a timely manner to permit the **Contractor** to carry out and complete the **Works** at the times specified in the **Contract**. If no such times are specified, then the **Purchaser** shall make such provision within a reasonable time with regard to any date set forth in Schedule 11 (Times of completion), the provisions of the **Approved Programme**, the actual progress of the **Works** and all other relevant circumstances.

4.2 If the **Contract** provides in Schedule 3 that the **Purchaser** shall carry out any work or provide any design, materials, facilities or services which are necessary to permit the carrying out and completion of the **Works** by the **Contractor** then, subject to any express descriptions thereof in the **Contract**, such work, design, materials, facilities or services shall be carried out or provided:

(a) with sound workmanship and materials, safely and in accordance with good engineering practice, applicable **Legislation** and codes;

(b) in a manner compatible with the proper carrying out and completion of the **Works** by the **Contractor**;

(c) at the times specified in the **Contract**, or if no such times are specified, then at reasonable times having regard to any date stated in Schedule 11 (Times of completion), the provisions of the **Approved Programme**, the actual progress of the **Works** and all other relevant circumstances.

4.3 If the **Purchaser** incurs costs which are to be reimbursed by the **Contractor** under the **Contract**, the **Purchaser** shall give notice to the **Contractor** of his intention to do so before incurring such costs and shall keep contemporary records of the work, materials and resources involved. Such records shall be open to inspection by the **Contract Manager** at all reasonable times.

4.4 The **Purchaser** shall provide personnel with adequate skills and experience to operate and maintain the **Plant** from the date of the **Take-Over Certificate** and shall make such personnel available for training in accordance with Schedule 9 (Training by Contractor).

4.5 The **Purchaser**, through the **Project Manager**, shall give the **Contractor** reasonable assistance in ascertaining the extent to which local laws or other restrictions could affect the execution of the **Works**.

4.6 The **Contractor** shall not be liable for the consequences of any errors or omissions in or arising from other work carried out by or on behalf of the **Purchaser**.

5. Decisions

5.1 Every **Decision**, notification, objection, claim or report to be given or made under the **Contract** shall be issued, given or made in writing. If any **Decision**, notification, objection, claim or report is issued, given or made orally, it shall be effective from the time and date when it is confirmed in writing.

5.2 Any challenge, where permitted by the **Contract**, to a **Decision**, objection, report or claim shall be supported by a written statement of grounds and a summary of material facts relating to each ground upon which it relies and shall be made within fourteen days of the receipt of the said **Decision**, objection, report or claim.

5.3 Minutes of meetings signed in accordance with Sub-clause 29.3 shall constitute a report in the context of this Clause and shall have the same effect as a written **Decision**.

6. Knowledge of Site conditions

6.1 If during the carrying out of the **Works** the **Contractor** encounters any physical condition on the **Site** which at the date of the Agreement could not reasonably have been foreseen by an experienced contractor possessed of all the information which the **Contractor** then had or could have obtained by a visual inspection of the **Site** or by reasonable enquiry, and if the **Contractor** considers that he will in consequence of such condition incur an increase in the time or cost of performing his obligations under the **Contract**, he shall give the **Project Manager** a notice under this Sub-clause within fourteen days of becoming aware of such unforeseen condition. Any such notice shall:

(a) expressly state that it is given under this Sub-clause;

(b) specify the condition encountered;

(c) specify the steps which the **Contractor** is taking or proposing to take to overcome the condition encountered;

(d) contain estimates of the effect on the **Approved Programme** of the condition and the additional cost which the **Contractor** is likely to incur; and

(e) state how the **Contractor** proposes to minimise such effect and additional cost and/or time.

If the **Contractor** is unable to incorporate in such notice any of the information listed in (b) to (e) above, he shall provide it to the **Project Manager** within a reasonable time.

6.2 The additional work of:

(a) complying with any instructions of the **Project Manager**, or

(b) taking proper and reasonable measures to overcome or deal with the physical conditions in the absence of any instruction from the **Project Manager**;

shall be deemed to be a **Variation** and the provisions of Clause 16 (Variations) shall apply.

7. Statutory and other obligations

See Guide Note E (Taxes)

7.1 The **Purchaser** and the **Contractor** shall in all matters relating to the **Contract** comply with all relevant **Legislation** and shall ensure that the **Plant** on completion and as proposed to be operated, as defined in the **Specification** or in any other contractual provision, complies therewith. The **Purchaser** shall obtain all permits required from government and local authority or any other permissions in connection with the use of the **Site** for construction, operation and maintenance of the **Plant** with the exception of those permissions, if any, specified in Schedule 1 (Description of the Works), the preparation and submission of the applications for which shall be the responsibility of the **Contractor**.

7.2 The **Contractor** shall develop, implement and maintain the Health and Safety Plan set out in Schedule 4 (Health and Safety) to take account of the development of the design, construction and maintenance procedures of the **Plant** having particular regard to the health and safety of all personnel involved with construction work on the **Site** and of all personnel who will be involved with the operation and maintenance of the **Plant.**

Such requirements shall not absolve the **Contractor** from any of his legal responsibilities for safety within the **Site**.

7.3 If after the date of the **Contractor's** tender for the **Works** as stated in the Agreement, there shall be enacted or brought into force any **Legislation** which causes an increase or decrease in the cost to the **Contractor** of carrying out any part of the **Works**, the amount of such increase or decrease shall be added to or deducted from the **Contract Price** and any fixed rate or charge stated in Schedule 18 (Cost elements, rates and charges), as the case may be, shall be adjusted accordingly. If such **Legislation** increases or decreases the time the **Contractor** requires to complete the **Works**, the **Approved Programme** shall be modified accordingly. Either party or the **Project Manager** shall give notice within fourteen days of becoming aware of any circumstance justifying such change or any modification to the **Approved Programme** and the **Project Manager** shall determine such change or modification as is reasonable by means of a **Variation Order**.

Any dispute as to whether or not the **Project Manager's Decision** is reasonable may be referred to an **Expert** in accordance with Clause 47 (Reference to an Expert).

7.4 Notwithstanding anything contained in this Clause, all taxes, levies, rates and charges assessed on the **Contractor**, including any changes therein and all taxes, withholdings and the like on or calculated by reference to the **Contractor's** profits or deemed profits, shall be borne by the **Contractor**. For the avoidance of doubt this Sub-clause does not apply to Value Added Tax, to the extent that provision is made for such tax in Sub-clause 39.10.

8. Patent and other protected rights

See Guide Note C (Intellectual property, know-how, confidentiality and information)

8.1 The **Purchaser** shall be responsible for the cost of all fees, royalties and other charges, if any, that may be payable through the **Contract** or otherwise under the terms of any licence or permission in respect of:

(a) any design of the **Plant** provided by the **Contractor**;

(b) the manufacture and supply of the **Plant**;

(c) any work done or method employed in the carrying out of the **Works**;

(d) any **Standard Software** or **Contractor's Software**; and

(e) operation or use of the **Plant**.

8.2 If any design of the **Plant** provided by the **Contractor**, or any item of the **Plant** manufactured or supplied by him, or any work done or method employed by him in the construction of the **Works** should infringe any patent, registered design, design right, trade mark, copyright or other intellectual property right protected by law, the **Contractor**, subject to Sub-clauses 8.3 and 8.4, shall indemnify the **Purchaser** against all losses, liabilities, cost and expenses that may result from such infringement, provided always that this indemnity shall not apply to any such patent, registered design, design right, trade mark, copyright or any other property rights protected by law first granted, registered or created after the date of tender as stated in the Agreement nor to any use of the **Plant** otherwise than for the purpose or in the manner indicated by or reasonably to be inferred from the **Contract**. The **Contractor** shall use reasonable endeavours to keep the **Project Manager** informed of any further patent, registered design, design right, trade mark or copyright which may be published, registered or created and which may affect the obligations of the parties hereunder.

8.3 In the event of any claim being made or proceedings instituted against the **Purchaser** to which the above indemnity applies, the **Purchaser** shall promptly notify the **Contractor** who shall:

(a) forthwith discuss with the **Purchaser** or the **Project Manager**, as required by the **Purchaser**, the action(s) that the **Contractor** intends to take in dealing with such claim and in conducting such proceedings;

(b) deal with such claim and conduct such proceedings in the **Purchaser's** name; and

(c) at all times keep the **Purchaser** or **Project Manager**, as the case may be, fully informed as to his progress in dealing with such claim or conducting such proceedings.

If the **Contractor** fails to notify the **Purchaser** within twenty-one days that he intends to deal with such claim or conduct such proceedings then the **Purchaser** shall be free to deal with such claim or conduct such proceedings on his own behalf. Unless the **Contractor** has failed to so notify the **Purchaser** within the period stated above, or fails at any time to comply with the other requirements of this Sub-clause, the **Purchaser** shall not make any admission prejudicial to such proceedings.

8.4 The **Purchaser** warrants that any **Documentation** or instructions furnished or given by him or by the **Project Manager** shall not cause the **Contractor** to infringe any patent, registered design, design right, trade mark, copyright or other intellectual property right protected by law in the performance of the **Contract** and shall indemnify the **Contractor** against all losses, liabilities, costs and expenses that may result from such infringement.

8.5 The copyright in all **Documentation** provided by the **Contractor** under the **Contract** shall remain vested in the **Contractor** or his **Subcontractors** as the case may be, subject to the pre-existing rights of any third party.

8.6 The **Purchaser** shall have the right to use such **Documentation** in connection with the design, construction, operation, maintenance and repair of the **Plant**. The **Purchaser** shall have the further right without additional payment to the **Contractor** to use any such **Documentation** for the purpose of making any improvement to or enlargement of the **Plant**, provided that such improvement or enlargement does not result in the construction of any separate or additional production plant or does not result in an increase in the production capacity of the **Plant** of more than twenty-five per cent over the original design.

8.7 Subject to the provisions of Clause 25 (Ownership of Materials) title in the **Applications Software** shall pass to the **Purchaser**. The **Contractor** shall procure on behalf of the **Purchaser** and grant to the **Purchaser** a right to the use of all forms of software for the lifetime of the **Plant**.

8.8 Within one month of the issue of the relevant **Take-Over Certificate**, the **Contractor** shall supply to the **Purchaser** a copy of the **Applications Software** codes, together with all **Documentation** necessary to allow the **Purchaser** to maintain, modify and compile the code into an executable form in combination with the **Standard Software** and the **Contractor's Software**, as detailed in the **Specification**.

9. Assignment and subcontracting

See Guidance on compiling Schedule 7 (Subcontracting) and Guide Note T (The Contracts (Rights of Third Parties) Act 1999)

9.1 Neither the **Purchaser** nor the **Contractor** shall without the previous consent of the other assign or transfer any benefit or obligation under the **Contract** to any other person in whole or in part, save that the **Contractor** may without such consent assign absolutely or by way of charge the right to receive any money which is or may become due to him under the **Contract**.

9.2 The **Contractor** shall not subcontract the whole of the **Works**. The **Contractor** shall agree with the **Project Manager** the extent to which part of the **Works** shall be subcontracted. The **Contractor** shall obtain the prior written approval of the **Project Manager** before entering into any subcontract. Except where a general approval has been given to a defined class of subcontracts, the approval given by the **Project Manager** shall relate only to the subcontract in respect of which it has been given.

The **Purchaser** shall have no obligation to reimburse the **Contractor** with respect to any subcontract for which prior written approval from the **Project Manager** has not been obtained. Subject to the provisions of Clause 10 (Nominated Subcontractors), the subcontracting by the **Contractor** of any part of the **Works** with the prior approval of the **Project Manager** shall not relieve the **Contractor** in any way whatsoever from his responsibility for the due performance of the **Contract**.

9.3 In the event that the **Contractor** or any **Affiliate** of the **Contractor** has any significant shareholding in any proposed **Subcontractor**, the **Contractor** shall so notify the **Project Manager** and any approval of a subcontract with any such **Subcontractor** given in the absence of such a notification shall have no effect.

9.4 Any subcontract entered into by the **Contractor** for the supply of any **Materials** or any other part of the **Works** shall so far as possible be negotiated in accordance with any procedure stated in Schedule 7 or be agreed with the **Project Manager**. In the absence of a stated or an agreed procedure the **Contractor** shall, if possible, obtain competitive fixed price tenders from qualified suppliers and shall submit them with his recommendations for selection to the **Project Manager** for approval.

9.5 The **Contractor** shall, in negotiating prices and terms with **Subcontractors**, ensure that any and all discounts, rebates and/or preferential terms that he is able to secure are passed on to the **Purchaser** and are not retained for the **Contractor's** benefit.

9.6 The **Contractor** shall ensure that each of his **Subcontractors** is bound to observe similar contractual obligations to those which bind him in these General Conditions and in any Special Conditions and to observe any provisions of the **Contract** which apply to subcontracts.

9.7 The **Contractor** shall use his best endeavours to include in every subcontract for the supply of **Materials**, a provision by which the **Subcontractor** shall undertake, to both the **Purchaser** and the **Contractor**, to make good all defects in the **Materials** arising out of manufacture, construction, workmanship or the **Subcontractor's** design, if any. Such provision shall provide for such liability to apply for a period of three hundred and sixty-five days from the date of any **Take-Over Certificate** or, if this is not possible, and with the agreement of the **Project Manager**, for a period of not less than five hundred and forty days from the date on which the respective **Materials** have been delivered to the **Site**, and shall also provide that the **Purchaser** shall be entitled to enforce its terms directly against the **Subcontractor**.

9.8 The **Contractor** shall ensure that any contract entered into with any **Subcontractor** shall include such terms and conditions that in the event of application of Clause 42 (Termination by the Purchaser for convenience) or Clause 43 (Termination for Contractor's default), the reasonable interest of both **Purchaser** and **Subcontractor** shall be safeguarded.

9.9 No subcontract shall contain any provision that makes payment to the **Subcontractor** by the **Contractor** dependent upon the receipt by the **Contractor** of any payment due under the **Contract**.

9.10 The **Contractor** shall include in any relevant subcontracts pursuant to which **Confidential Information** referred to in Sub-clause 20.2 is to be disclosed to the **Subcontractor** a clause or clauses providing that the **Subcontractor** shall observe the same obligations of confidentiality as those of the **Contractor** hereunder, and that the **Purchaser** shall be entitled to enforce such clause or clauses directly against the **Subcontractor**.

9.11 The **Purchaser** hereby agrees that the obligations of the **Purchaser** under Sub-clause 20.3 shall be capable of being enforced directly by any **Subcontractor** who has provided such information to the **Contractor**.

9.12 Save as expressly provided in this Clause, no person other than a party to this **Contract** shall have any right to enforce any of its terms under the Contracts (Rights of Third Parties) Act 1999.

10. Nominated Subcontractors

See Guidance on compiling Schedule 7 (Subcontracting)

10.1 If the **Contract** provides that any work shall be done or **Materials** or services supplied by a **Subcontractor** named in the **Contract** and stated to be a **Subcontractor** nominated by the **Purchaser**, then Sub-clauses 10.6 and 10.7 shall apply in respect of that nominated **Subcontractor**.

10.2 If the **Contract** provides, whether or not by means of a prime cost sum, that certain work shall be done or **Materials** or services supplied by a **Subcontractor** nominated by the **Project Manager**, then the **Project Manager** shall make the necessary nomination at a reasonable time having regard to the progress of the **Works** and, subject to the provisions of Sub-clauses 10.3 and 10.4, the **Contractor** shall enter into a subcontract in accordance with such nomination and Sub-clauses 10.6 and 10.7 shall apply.

10.3 Upon receipt of a nomination pursuant to Sub-clause 10.2, the **Contractor** shall within twenty-eight days of such nomination and before entering into such subcontract give notice to the **Project Manager** with reasons if he considers that:

(a) compliance with the nomination would prevent the **Contractor** from or hinder him in fulfilling any of his obligations under the **Contract**; or

(b) the nominated **Subcontractor** is unwilling to enter into a subcontract with the **Contractor** on terms which are compatible with the **Contract** and provide the **Contractor** with reasonable remedies in the event of a breach thereof by such **Subcontractor**; or

(c) the nominated **Subcontractor** is unlikely to be reliable or competent in his performance of the subcontract.

10.4 Upon receipt of any notification under the preceding Sub-clause 10.3, the **Project Manager** shall unless otherwise agreed:

(a) withdraw the nomination and nominate another **Subcontractor** in accordance with Sub-clause 10.2; or

(b) withdraw the nomination and issue a **Variation Order** omitting the work, **Materials** or services in question, in which case the **Purchaser** may enter into a direct contract with the nominated **Subcontractor**; or

(c) issue a **Variation Order** confirming the nomination, in which case the **Contractor** shall enter into a subcontract with the nominated **Subcontractor** on terms agreed with the **Project Manager**.

10.5 In the event that the **Contractor** does not give notice to the **Project Manager** under Sub-clause 10.3 above, he shall be obliged to accept the **Project Manager's** nomination.

10.6 If any payment to a nominated **Subcontractor** has been improperly withheld by the **Contractor** then the **Purchaser** shall be entitled to give notice in writing to the **Contractor** that he intends to make such payment direct to the **Subcontractor**. If after fourteen days, the payment is still withheld and the **Contractor** has failed to furnish adequate reason why it should be withheld, the **Purchaser** may, except in the event of the **Contractor's** insolvency, make such payment and shall be entitled to withhold from the **Contractor** the amount so paid.

10.7 To the extent that any additional cost or expense incurred by the **Contractor** as a result of the failure of any nominated **Subcontractor** to perform his obligations under the relevant subcontract is not recovered by the **Contractor** from the **Subcontractor**, the **Purchaser** shall pay the **Contractor** such additional cost or expense, provided that:

(a) the **Contractor** shall give notice to the **Project Manager** forthwith of such failure;

(b) the nominated **Subcontractor's** failure to perform did not result from any failure of the **Contractor** to meet his obligations under the **Contract**.

11. The Project Manager

11.1 The **Project Manager** shall have full authority to act on behalf of the **Purchaser** in connection with the **Contract**, except to the extent otherwise stated in Sub-clause 36.6, and:

(a) the **Purchaser** shall cause the **Project Manager** to perform in a timely manner every act required under the **Contract** to be performed by the **Project Manager**;

(b) any obligation stated under the **Contract** to be an obligation of the **Project Manager** shall be deemed to be an obligation of the **Purchaser**;

(c) the **Purchaser** shall be responsible for any act, neglect or omission of the **Project Manager** as if it were an act, neglect or omission of the **Purchaser**; and

(d) in all matters where the **Project Manager** is required or authorised under the **Contract** to exercise his discretion or make a judgment or form an opinion, he shall do so to the best of his skill and judgment as a professional engineer and shall be impartial between the **Purchaser** and **Contractor**.

11.2 The **Project Manager** may at any time instruct the **Contractor** to execute the **Works** or any part thereof as the **Project Manager** may require and the **Contractor** shall comply with such instruction within a reasonable period to be agreed between the **Project Manager** and the **Contractor**.

If the **Contractor** is of the opinion that compliance with an instruction of the **Project Manager** would:

(a) constitute a change in the type or extent of the **Works** or services to be supplied by the **Contractor** under the **Contract**;

(b) delay completion of the **Works**; or

(c) prevent the **Contractor** from fulfilling his obligations under the **Contract**;

he shall so inform the **Project Manager** in writing within seven days. The **Project Manager** may request the **Contractor** to provide justification for his opinion. The cost incurred by the **Contractor** in preparing the justification shall form part of the **Contract Price**. If the **Project Manager** accepts the **Contractor's** opinion, he shall withdraw the instruction and, if he wishes, substitute a **Variation Order** under the provisions of Clause 16 (Variations).

11.3 The **Purchaser** may from time to time appoint some other individual as **Project Manager** in place of the individual previously so named or appointed and shall give notice to the **Contractor** without delay.

11.4 If the individual initially named in the **Contract** is not an employee of the **Purchaser** or any **Affiliate** of the **Purchaser**, then the **Purchaser** shall not thereafter have the right to appoint as **Project Manager** any individual who is an employee of the **Purchaser** or any **Affiliate** of the **Purchaser**, unless the **Contractor** consents.

11.5 The **Project Manager** may by notice to the **Contractor** appoint any individual to act as a **Project Manager's Representative** and may from time to time by further notice change such appointment. Any such **Project Manager's Representative** shall have authority to condemn any designs, workmanship or materials which are not in accordance with the **Contract**, but the **Contractor** may within seven days of such condemnation and before complying therewith appeal against it to the **Project Manager**. Any such **Project Manager's Representative** shall also have authority to receive on behalf of the **Project Manager** any notice to be given to the **Project Manager** under the **Contract** by the **Contractor**, but the **Contractor** shall also send a copy of any such notice directly to the **Project Manager**.

11.6 Subject to the provisions of Sub-clauses 12.5 and 28.4, the **Project Manager** may by notice to the **Contractor** authorise any **Project Manager's Representative** to exercise any of the powers and functions of the **Project Manager** under the **Contract** and may by further notice cancel or modify any such authority, but not with retrospective effect.

11.7 If in the absence of the necessary authority from the **Project Manager** a **Project Manager's Representative** has given the **Contractor** any **Decision** which should have been given by the **Project Manager**, the **Contractor** may, within seven days of receipt, advise the **Project Manager** of such **Decision**. Unless the **Project Manager** rescinds the **Decision** within seven days of receipt of such advice, the said **Decision** shall have effect as if it had been given by the **Project Manager**. In the event that the **Contractor** does not advise the **Project Manager** of any such **Decision** within seven days of receipt, the **Decision** shall be conclusively deemed to have been given within the authority of the **Project Manager**, provided that it was given in writing.

11.8 Save as otherwise provided in this **Contract**, all communications between the **Purchaser** and the **Contractor** relating to the **Works** shall be between the **Project Manager** and the **Contract Manager**. No communication by any other means shall be of any effect unless receipt thereof is acknowledged by the **Project Manager** or **Contract Manager** as the case may be.

12. Contract Manager and Contractor's staff

See Guidance on compiling Schedule 8 (Contractor's named personnel)

12.1 The **Contract Manager** shall have full authority to act on the **Contractor's** behalf in connection with the **Contract**. The **Contractor** shall not replace the **Contract Manager** without the previous consent of the **Project Manager**. The **Contractor** shall at all times until the issue of the last **Final Certificate** ensure that a suitable individual is appointed to act as **Contract Manager**. The **Contract Manager** may appoint a deputy to act in his place in his absence. The **Contract Manager** shall inform the **Project Manager** in writing of his intention to appoint such a deputy who will be deemed to have full authority to act on the **Contractor's** behalf.

12.2 From the commencement of work at the **Site** until the whole of the **Plant** has been **Accepted** by the **Purchaser**, the **Contractor** shall ensure that some suitable individual is employed at the **Site** as **Site Manager**. The **Contractor** shall notify the **Project Manager** of the individual so appointed as **Site Manager** and shall not make or change such appointment without the previous consent or requirement of the **Project Manager**.

The **Site Manager** or his deputy shall supervise all work done at the **Site** by the **Contractor** and shall receive on behalf of the **Contractor** all **Decisions** given to the **Contractor** at the **Site** by the **Project Manager** or any **Project Manager's Representative**. The **Site Manager** shall be present at the **Site** throughout normal working hours except when on leave, sick or absent for reasons connected with the proper performance of the **Contract** or as agreed with the **Project Manager**. Whenever the **Site Manager** is absent from the **Site** during normal working hours, a suitable individual shall be appointed to act as his deputy. The **Site Manager** shall not be employed by the **Contractor** to do work under contracts with others without the prior written agreement of the **Project Manager**.

12.3 The **Contractor** shall provide the further supervisory staff specified in Schedule 8. Without prejudice to the foregoing, the **Contractor** shall ensure that there are at all times at the **Site** sufficient suitably qualified and experienced staff to supervise all work being done by the **Contractor** at the **Site** and, where required by the **Contract**, to advise and assist the **Purchaser** in starting up the **Plant**, carrying out the performance tests, if any, and operating the **Plant** until its **Acceptance**.

12.4 The **Contractor** shall use all reasonable endeavours to ensure that any key personnel named in Schedule 8 shall continue to be employed in their specified capacities on the work for so long as and to the extent that the **Works** require. The **Contractor** shall not, without the written consent of the **Project Manager**, replace any such key personnel.

12.5 If the **Project Manager** is of the opinion that the **Site Manager** or any member of the **Contractor's** supervisory staff at the **Site** is incompetent or has been guilty of misconduct or serious breach of his duties, he may by notice to the **Contractor** require such person to leave the **Site** with immediate effect. The **Contractor** shall be under an obligation to act on the **Project Manager's** notice forthwith and to replace such person as soon as practicable. Any such notice given by the **Project Manager** shall be given in good faith and shall be final, binding and conclusive and not capable of being reviewed, revised or reversed by resolution under Clause 45 (Disputes) or by an Adjudicator under Clause 46 (Adjudication) or reference to an **Expert** under Clause 47 (Reference to an Expert) or under Clause 48 (Arbitration) or court proceedings. The **Project Manager** shall not delegate the power to give such notice to any **Project Manager's Representative**. The **Purchaser** shall have no obligation to reimburse the **Contractor** the cost of replacing such person.

13. Times of completion and Approved Programme

See Guidance on compiling Schedule 11 (Times of completion) and Guide Note S (Completion, taking over, testing and start-up)

13.1 Subject to the provisions of Clause 14 (Delays), the **Contractor** shall complete the construction of the **Plant** ready for the carrying out of the take-over procedures on or before the date, or within the period, specified in Schedule 11 and shall also complete any specified section of the **Plant** and do any other thing in the performance of the **Contract** on or before the dates, or within the periods, specified in the said Schedule.

13.2 If Schedule 11 provides for events the occurrence of which can affect the amounts paid by one party to the other, the **Contractor** shall provide evidence that each such event has taken place and the **Project Manager** shall, if he is satisfied with such evidence, thereupon issue a suitable certificate for each event stating the date on which it occurred.

13.3 The **Contractor** shall, within the time stated in Schedule 11, prepare and submit to the **Project Manager** for his approval a programme of work setting out in such manner as the **Project Manager** may reasonably require the sequence in which and dates by which the **Contractor** proposes to perform his obligations under the **Contract** and the date(s) by which the **Contractor** reasonably requires that the **Purchaser** should provide any further **Documentation** or information or take any other action to permit the **Contractor** so to perform his obligations. If the programme submitted by the **Contractor** accords with any dates and periods specified in the **Contract** and is otherwise reasonable, the **Project Manager** shall approve it and it shall be the **Approved Programme**.

The **Contractor** shall use his reasonable endeavours to perform his obligations under the **Contract** in accordance with the provisions of the **Approved Programme**.

13.4 The **Contractor** shall submit details of the qualifications, experience and numbers of the personnel whom he proposes to employ on the **Works** at various times, and the other resources which he will have available in order to perform his obligations under the **Contract** and in accordance with the programme. Such details shall be in such form as the **Project Manager** may require and shall be submitted to him with the programme of work. The **Contractor** shall employ and make available personnel and resources conforming to such proposals.

13.5 If at any time the performance of the **Contract** falls behind the **Approved Programme**, or it becomes clear that it will so fall behind, then the **Project Manager** may require the **Contractor** either to take such steps as may be practicable in order to achieve the **Approved Programme** or to revise the **Approved Programme** in the light of the circumstances and to re-submit it to him for his approval. If the **Project Manager** approves the revised programme it shall thereafter be the **Approved Programme**.

13.6 Without prejudice to Sub-clause 13.5, if the **Project Manager** decides that the rate of progress by the **Contractor** in carrying out the **Works** is likely to prejudice the **Contractor's** ability to complete the construction of the **Plant**, or any specified section thereof, in accordance with the provisions of Sub-clause 13.1, and that this is due to a cause for which the **Contractor** is responsible, the **Project Manager** may give notice to that effect to the **Contractor**. Following such notice the **Contractor** and the **Project Manager** shall agree a course of action, including the use of additional resources, to remedy the potential delay, and the **Contractor** shall use his best endeavours to implement the agreed course of action.

13.7 If when the **Project Manager** requires the **Contractor** to revise the **Approved Programme**, the **Contractor** fails to submit such revised programme within a reasonable time, or if the **Contractor** submits a revised programme which the **Project Manager** is unable to approve for good reason, the **Project Manager** may instruct the **Contractor** in writing to make reasonable revisions to the programme, and the **Contractor** shall forthwith make such revisions and the revised programme shall thereafter be the **Approved Programme**.

13.8 The exercise by the **Project Manager** of his powers under Sub-clauses 13.5, 13.6 and 13.7 above shall not affect any of the **Contractor's** obligations to the **Purchaser** under the **Contract**.

14. Delays

14.1 If the **Contractor** is delayed in the performance of any of his obligations under the **Contract** by any of the matters specified below, or if either party is delayed by Force Majeure in the performance of any of his obligations under the **Contract**, the relevant party shall forthwith give notice to the **Project Manager** and as appropriate to the **Contractor**.

As soon as reasonably possible, the **Contractor** shall advise the **Project Manager** of the extension of any date or period specified in the **Contract** for the completion of such obligations which he considers would be fair and reasonable in the circumstances. The **Contractor** shall keep contemporaneous records of the circumstances, extent and effect of such delay. The **Project Manager** shall, within fourteen days of the time that the extent and consequences of any such delay are known, issue a **Variation Order** both to the **Purchaser** and to the **Contractor** stating the appropriate extension to the **Approved Programme** and to Schedule 11 (Times of completion) or, if appropriate, to the period in Schedule 16 (Performance tests and procedures) by the end of which the **Plant** should have passed all its performance tests. If either party does not agree with such extension, then the matter may be referred to an **Expert** in accordance with Clause 47 (Reference to an Expert).

The matters entitling the **Contractor** to an extension under this Sub-clause are delays caused by:

(a) the occurrence of conditions to which the provisions of Sub-clause 6.1 apply;

(b) a **Variation** ordered by the **Project Manager** (other than a **Variation** ordered by reason of the **Contractor's** default) except where the delay is already covered in a **Variation Order** issued by the **Project Manager** under Sub-clause 16.3;

(c) the giving of any **Suspension Order** by the **Project Manager**, except where given by reason of the **Contractor's** default;

(d) a breach of the **Contract** by the **Purchaser**; or

(e) the failure of any **Subcontractor** nominated by the **Project Manager** in accordance with Clause 10 (Nominated Subcontractors) to perform such **Subcontractor's** obligations despite all due supervision by the **Contractor**.

14.2 In the context of this Clause, 'Force Majeure' shall mean any circumstance beyond the reasonable control of either party which prevents or impedes the due performance of the **Contract** by that party including, but not limited to, the following:

(a) government action or trade embargo;

(b) war, hostilities or acts of terrorism;

(c) riot or civil commotion;

(d) epidemic;

(e) earthquake, flood, fire or other natural physical disaster;

(f) exceptionally severe weather conditions or the consequences thereof;

(g) denial of the use of any railway, port, airport, shipping service or other means of public transport; or

(h) industrial disputes, other than any solely confined to the **Contractor** and/or his **Subcontractors** or their employees including employees of any **Affiliate** of the **Contractor** or **Subcontractor**.

The mere shortage of labour, materials or utilities shall not constitute Force Majeure unless caused by circumstances which are themselves Force Majeure.

14.3 Both parties shall at all times use all reasonable endeavours to minimise any delay in the performance of their obligations under the **Contract**, whatever may be the cause of such delay.

14.4 In the event that either party is delayed by Force Majeure, each party shall bear his own costs of such delay, other than the **Contractor's** costs incurred at the **Site** which shall form part of the **Contract Price**.

14.5 If performance of the **Works** is substantially prevented by Force Majeure for a continuous period of one hundred and twenty days, either party may terminate the employment of the **Contractor** by written notice to the other. In the event of such termination the rights and obligations of the parties shall be the same as if the employment of the **Contractor** had been terminated by the **Purchaser** under Clause 42 (Termination by the Purchaser for convenience).

15. Damages for delay

See Guidance on compiling Schedule 12 (Liquidated damages for delay)

15.1 If the **Contractor** fails to complete the **Plant** or any specified section thereof or to do any other thing in accordance with Schedule 11 (Times of completion), the **Contractor** shall pay the **Purchaser** liquidated damages as prescribed in Schedule 12, but shall have no liability to pay damages in excess of the maximum (if any) stated in Schedule 12.

15.2 If after liquidated damages for delay have become payable in respect of any part of the **Plant** the **Project Manager** issues a **Variation Order** or a physical condition is encountered as envisaged in Sub-clause 6.1, either of which delays the **Contractor** and in the opinion of the **Project Manager** properly entitles the **Contractor** to an extension of time in respect of such further delay to that part of the **Plant**, the **Project Manager** shall forthwith so inform the **Contractor** and the **Purchaser** in writing.

The **Purchaser's** further entitlement to liquidated damages in respect of that part of the **Plant** shall thereupon be suspended until the **Project Manager** notifies the **Contractor** and **Purchaser** in writing that such further delay has come to an end.

Such suspension shall not invalidate any entitlement to liquidated damages which accrued before the period of further delay started to run and (subject to any final review of the circumstances) any monies already deducted or paid as liquidated damages for delay may be retained by the **Purchaser**.

15.3 Any dispute arising under this Clause may be referred to an **Expert** in accordance with Clause 47 (Reference to an Expert).

16. Variations

See Guide Note I (Variations)

16.1 A **Variation** shall mean any alteration to the **Plant**, method of working, programme of work or to the type or extent of the **Works**, which is an amendment, omission or addition thereto (other than any amendment, omission, or addition which is necessary for the **Plant** to comply with the **Specification**). A closer definition, a minor change in detail or an immaterial alteration in the manner in which the work is to be performed shall not constitute a **Variation**. The **Contractor** shall make no **Variation** except as ordered in writing by the **Project Manager**. If no **Variation Order** has been issued by the **Project Manager** in respect of an amendment, omission or addition which the **Contractor** considers is significant by itself or when taken together with such previous changes not themselves covered by any **Variation Order**, the **Contractor** may give notice to the **Project Manager** that before complying with such amendment, omission or addition the **Contractor** requires a **Variation Order** to be issued.

16.2 At any time during the performance of the **Contract** the **Project Manager** may give the **Contractor** a **Variation Order**, specifying the **Variation** desired and, subject to the provisions of this Clause, the **Contractor** shall be bound as if the **Variation Order** were part of the original **Contract**.

16.3 If the circumstances so justify, a **Variation Order** issued by the **Project Manager** under this Clause shall amend the **Approved Programme** and any of the dates set forth in Schedule 11 (Times of completion).

16.4 The **Project Manager** may at any time instruct the **Contractor** to prepare, or to assist him in the preparation of, a potential **Variation** and the **Contractor** shall comply with such instruction and furnish to the **Project Manager** his proposals for the form, scope and probable cost of the **Variation** at such time and in such detail as the **Project Manager** shall require.

16.5 Notwithstanding that the cost of complying with the **Variation Order** remains to be estimated, the **Project Manager** may issue such **Variation Order** pursuant to Sub-clause 16.2 if he feels that delaying the **Variation Order** pending such estimate would unnecessarily prejudice the satisfactory completion of the **Works** or cause avoidable harm to the interests of the **Purchaser**. In all other cases the **Project Manager** shall afford the **Contractor** reasonable opportunity to comment upon any **Variation** he proposes to order and shall request the **Contractor** to set out his proposals for consequent adjustments to the **Works**. The **Contractor** shall provide his proposals within fourteen days of receipt of such request or within such longer period as the **Project Manager** may agree.

16.6 If the **Contractor** is of the opinion that compliance with any **Variation Order** would prevent him from or hinder him in fulfilling any obligation under the **Contract**, he shall so notify the **Project Manager** within seven days giving reasons why he considers he would be so prevented or hindered. No **Variation Order** in respect of which such a notification is given by the **Contractor** shall become binding unless it is thereafter confirmed by the **Project Manager**. If the **Variation Order** is so confirmed then the obligations of the **Contractor** shall thereupon be modified to such extent as the **Contractor** and the **Project Manager** may agree.

16.7 The **Contractor** may also object to any **Variation** ordered or proposed upon the grounds that compliance therewith would:

(a) when combined with all **Variations** previously ordered increase or decrease the first agreed estimate of the **Contract Price** by more than twenty-five per cent;

(b) in the case of a **Variation Order** issued after the date of a **Take-Over Certificate** for all or part of the **Works**, increase the first agreed estimate of the **Contract Price** by more than five per cent;

(c) require the **Contractor** to act in breach of any enforceable undertaking or agreement with a third party;

(d) cause him to infringe any legal obligation, patent, registered design, design right, copyright or other protected right of any third party;

(e) require the **Contractor** to do work or to exercise skills which are not of the kind the **Contractor** undertakes in the ordinary course of his business, unless the **Variation Order** specifies that the work shall be performed or those skills exercised by a **Subcontractor** pursuant to Clause 10 (Nominated Subcontractors); or

(f) involve the employment of personnel resources beyond the current capacity of the **Contractor** where he can demonstrate that services of additional appropriately qualified personnel cannot reasonably be obtained.

Any such objection by the **Contractor** shall be made in writing to the **Project Manager** as soon as reasonably practicable and not later than fourteen days from receipt of the order. In such case, the **Project Manager** shall notify the **Contractor** either that he withdraw the **Variation Order** or proposal, or that he does not accept the validity of the **Contractor's** objection, in which case any dispute may be referred to an **Expert** in accordance with Clause 47 (Reference to an Expert). The decision of the **Contractor** not to object to any **Variation** ordered or proposed hereunder shall not affect his right to object to any subsequent **Variation Order** or proposal.

16.8 The **Contractor's** costs in preparing a potential **Variation** in accordance with Sub-clause 16.4, or of commenting upon a proposed **Variation** in accordance with Sub-clause 16.5, or of responding to a **Variation Order** in accordance with Sub-clause 16.6 or 16.7, shall form part of the **Contract Price** whether or not any such potential or proposed **Variation** is ordered or amended or such **Variation Order** is confirmed or withdrawn. The provisions of Sub-clause 19.1 shall apply.

16.9 If a disagreement arises between the **Project Manager** and the **Contractor** about any cost incurred by the **Contractor** pursuant to this Clause or Clause 17 (Contractor's Variations) or any modification to the obligations of the **Contractor** necessary to incorporate into the **Works** or the **Approved Programme** or any **Variation** that has been ordered by the **Project Manager**, the dispute may be referred to an **Expert** in accordance with Clause 47 (Reference to an Expert).

16.10 The **Contractor** shall establish and maintain contemporary records of the work, **Materials** and resources required to undertake any **Variation** and their cost together with such additional records as the **Project Manager** may reasonably direct at the time of issuing the **Variation Order**, and all such records shall be open to inspection by the **Project Manager**.

17. Contractor's Variations

See Guide Note I (Variations)

17.1 The **Contractor** may at any time during the performance of the **Contract** submit a proposal for a **Variation** to the **Project Manager**. Initially such proposal should be given as a brief written outline as soon as the **Contractor** is first aware of the possibility of the **Variation**.

17.2 If the **Project Manager** wishes to give the proposal further consideration he shall agree with the **Contractor** the extent of the detail that he requires the **Contractor** to prepare in substantiating the proposal or in developing full details of the proposed **Variation**, and thereafter the **Contractor** shall undertake such further preparatory work. The provisions of Sub-clause 19.1 shall apply to such further preparatory work and the cost thereof shall form part of the **Contract Price**.

17.3 If the **Project Manager** decides that the proposed **Variation** should be incorporated into the **Works** or the **Approved Programme** he shall so order and the provisions of Clause 16 (Variations) shall apply thereto. The **Project Manager** may instead order or propose an alternative **Variation** in accordance with Clause 16.

The **Project Manager's Decision** on the proposal for a **Variation** or on the detailed proposed **Variation** under Sub-clauses 17.1 and 17.2 respectively shall be given within fourteen days, or such longer period as the **Project Manager** notifies the **Contractor** within the said fourteen days is reasonably required for the **Project Manager** and the **Purchaser** to consider the proposal or proposed **Variation**. Other than in the special circumstances described in Sub-clause 17.4, **Decisions** of the **Project Manager** not to order or propose a **Variation** shall be final, binding and conclusive and not capable of being reviewed, revised or reversed by an Adjudicator under Clause 46 (Reference to Adjudication) or by reference to an **Expert** under Clause 47 (Reference to an Expert) or under Clause 48 (Arbitration) or court proceedings.

17.4 The special circumstances referred to in Sub-clause 17.3 are those where the **Contractor** has stated in his proposals that the object of the **Variation** is to eliminate a potential defect in the **Works** or a specific hazard to any person or property in the performance of the **Works** or in the operation or use of the **Plant** (including any breach of a duty imposed by any applicable health and safety legislation). The **Project Manager's Decision** not to order the **Variation** or not to order or propose an alternative **Variation** shall be notified to the **Contractor**. If the **Contractor** shall dispute the **Project Manager's Decision**, the dispute may be referred to an **Expert** in accordance with Clause 47.

17.5 Without prejudice to the **Contractor's** obligations under Clause 21 (Documentation) and Clause 37 (Liability for Defects), if the **Contractor** shall become aware that any item of the **Plant** or any part of the **Works** to be performed by the **Contractor** has been incorrectly specified in the **Contract**, the **Contractor** shall immediately give notice to the **Project Manager**. The **Project Manager** and the **Contractor** shall as soon as possible meet to consider what action and what **Variation** may be needed.

18. Contract Price

See Schedules 18 (Cost elements, rates and charges) and 19 (Terms of payment)

18.1 The **Purchaser** shall make payments to the **Contractor** in accordance with the **Contract** for carrying out the **Works** and the total of all such sums payable shall be the **Contract Price**.

18.2 If in performing the **Contract** the **Contractor** may need to provide **Materials** and/or services for which the amount payable cannot be calculated from Schedules 18 and 19, then the **Project Manager** and the **Contractor** shall agree the amounts to be included in the **Contract Price** for such items and the aforesaid Schedules shall be amended accordingly.

18.3 If the **Project Manager** and the **Contractor** cannot agree on the amount of time or other measure of any resources, personnel, **Contractor's Equipment** or **Materials** used by the **Contractor** in undertaking the **Works** or on any amendment to Schedules 18 and 19 in accordance with Sub-clause 18.2, the dispute may be referred to an **Expert** in accordance with Clause 47 (Reference to an Expert).

19. Records and audits

19.1 In addition to his obligations under Sub-clause 3.7, the **Contractor** shall keep books and records generally in accordance with the provisions of Schedule 20 (Commercial records) and also to the **Project Manager's** satisfaction to enable him to carry out effective control with respect to all payments made or due to be made pursuant to the **Contract**, other than in respect of the make-up of any fixed or unit rate or price.

19.2 The **Purchaser** shall have the right at any time to carry out audits of such books and records and any other related documents and have such audits carried out by any auditing firm appointed by the **Purchaser** until three hundred and sixty five days after the date of the last **Final Certificate**.

20. Confidentiality

See Guide Note C (Intellectual property, know-how, confidentiality and information)

20.1 '**Confidential Information**' shall mean all **Documentation** and other technical or commercial information in any form obtained directly or indirectly from the **Purchaser** or the **Project Manager** by the **Contractor**, or from the **Contractor** by the **Purchaser** or the **Project Manager**, or which is generated by the **Contractor** or any **Subcontractor** in connection with the **Contract**, other than information:

(a) which is or becomes generally available in the public domain other than by any unauthorised actions of either of the parties to this **Contract**;

(b) which is or comes into the possession of one party other than in breach of a duty of confidence to the other party; or

(c) which is or comes into the possession of the **Contractor** with the right to disclose.

20.2 The **Contractor** shall not, without the previous written consent of the **Purchaser**, use, publish or disclose to any person, nor cause nor permit any of his **Affiliates**, employees, agents or **Subcontractors** to use, publish or disclose any **Confidential Information** otherwise than for the performance of his duties under the **Contract**. If required by the **Purchaser**, the **Contractor** shall cause any of his **Subcontractors** and/or any of their respective **Affiliates**, employees or agents to enter into an individual written obligation to the **Purchaser** to comply with the provisions of this Sub-clause.

20.3 Except as permitted in Sub-clause 8.6, the **Purchaser** shall not, without the previous written consent of the **Contractor**, use, nor cause nor permit any of his **Affiliates**, employees or agents to use any **Confidential Information** received by the **Purchaser** otherwise than for the design, construction, operation or maintenance of the **Plant** and associated facilities.

20.4 The **Contractor** shall not, other than for the purposes of performing the **Works**, take or permit to be taken any photograph or other image of the whole or any part of the **Plant** or any other property of the **Purchaser**, or any physical or virtual model thereof, without the prior written consent of the **Purchaser** or the **Project Manager**. Any such photograph or other image shall be regarded as **Confidential Information** within the terms of Sub-clause 20.1. No photograph or other image so taken shall be used for the purposes of publicity without the prior written consent of the **Purchaser**.

20.5 The provisions of this Clause shall survive and remain in full force following the completion of the **Works**.

21. Documentation

See Guidance on compiling Schedule 2 (Documentation)

21.1 The **Contractor** shall appoint competent persons who shall be responsible for the checking and approval of all **Documentation** provided by the **Contractor**. No **Documentation** shall be both checked and approved by the same individual. If such persons are not named in the **Contract**, the **Contractor** shall forthwith make such appointments and shall notify the **Project Manager** accordingly.

21.2 The **Contractor** shall submit to the **Project Manager** for his approval the **Documentation** listed in Schedule 2 under the heading 'Documentation for approval' at the time or times stated in Schedule 11 (Times of completion) or in the **Approved Programme**.

Within fourteen days of the receipt by him of any **Documentation** for his approval, the **Project Manager** shall return one copy thereof to the **Contractor** with his approval, conditional approval or rejection endorsed thereon. If the **Project Manager** fails to take any such action within the said fourteen days, the **Documentation** shall be deemed to have been approved by the **Project Manager** and the **Contractor** shall notify the **Project Manager** accordingly.

21.3 The **Project Manager's** approval of any **Documentation** submitted to him shall not relieve the **Contractor** of any of his responsibilities under the **Contract**. The **Contractor** shall not depart from any approved **Documentation** unless he has first submitted amended **Documentation** to the **Project Manager** and obtained his approval thereof.

21.4 The **Contractor** shall not challenge the disapproval by the **Project Manager** of any **Documentation** or amended **Documentation** except on the grounds that the requirement of the **Project Manager** is not in accordance with some express provision of the **Contract** or that it is contrary to good engineering practice or is likely to have an adverse effect on health or safety. If the **Contractor** does not accept the disapproval by the **Project Manager** of any **Documentation**, the dispute may be referred to an **Expert** in accordance with Clause 47 (Reference to an Expert).

21.5 In each case where the **Project Manager** gives conditional approval of any **Documentation** subject to any comment or query he may have thereon, he shall make such comment or query in writing on or attached to such **Documentation**, and the **Contractor** shall re-submit it with his response to the comment or query. The **Contractor** shall proceed on the basis of the conditional approval and the **Project Manager** will approve such **Documentation** without conditions when it is re-submitted.

21.6 The **Project Manager** shall have the right at any time on reasonable notice to examine any **Documentation** which has been or is being prepared by the **Contractor** or his **Subcontractor**s for the purposes of the **Contract** except any **Documentation** of a class or description which the Special Conditions or Schedule 2 state shall not be shown to the **Project Manager**.

21.7 Schedule 2 may include a description of **Documentation** under a heading 'Documentation for information only' of which the **Project Manager** requires to examine copies, in addition to those to be submitted under Sub-clause 21.2. If specified in Schedule 2, the **Contractor** shall submit such copies to the **Project Manager** as soon as each becomes available in the form specified in this Schedule.

21.8 **Documentation** relating to any items subject to statutory design requirements and/or insurance approval shall be made available to the **Purchaser** and/or his insurers or agents at times appropriate for such purposes, with due regard to the **Approved Programme**.

21.9 Before the **Plant** is taken over pursuant to Clause 33 (Taking over), the **Contractor** shall supply the **Documentation** and manuals specified in Schedule 2 under the heading 'Final Documentation and Manuals' for the **Plant** as actually constructed.

21.10 The **Project Manager** shall review the **Documentation** supplied by the **Contractor** in accordance with Sub-clauses 21.8 and 21.9 for completeness and accuracy, and may require its amendment by the **Contractor** as necessary. The provisions of Sub-clause 21.14 shall apply to the **Contractor's** cost of such amendment.

21.11 Whenever the **Contractor** makes any change to the **Plant** as represented in the final documentation and manuals, he shall, within a period of sixty days, provide the **Purchaser** with new **Documentation** revised to take account of such change.

21.12 The **Contractor** shall promptly correct any error, discrepancy or omission in any **Documentation** prepared by him or on his behalf. The provisions of Sub-clause 21.14 shall apply to the **Contractor's** cost of such correction.

21.13 The **Contractor** shall if required by the **Project Manager** correct any error, discrepancy or omission in any **Documentation** provided to him by the **Purchaser** or the **Project Manager** for the purposes of the **Works**. If the need for such correction has arisen by reason of any inaccurate data, **Documentation** or information provided by the **Contractor** and on which the **Purchaser** or the **Project Manager** was entitled to and did reasonably rely, the provisions of Sub-clause 21.14 shall apply to the **Contractor's** cost of such correction. Otherwise the **Contractor's** cost shall form part of the **Contract Price**.

21.14 The **Contractor's** cost of any amendment or correction under Sub-clause 21.10, 21.12 or 21.13 shall form part of the **Contract Price**, except where:

(a) the inaccurate data, **Documentation** or information was provided to the **Contractor** by a **Subcontractor** who makes good the error, discrepancy or omission without further charge to the **Contractor**, or the cost of the correction work is paid by the **Subcontractor** to the **Contractor**;

(b) the **Contractor's** work of preparing the **Documentation** is to be paid for by means of a fixed rate charge or fee; or

(c) the error, discrepancy or omission in the **Documentation** is due to a failure of the **Contractor**, his employees or agents to exercise the requisite degree of skill and care which would ordinarily be exercised by properly qualified and competent persons engaged in the same or a similar business to that of the **Contractor**.

22. Inspection and pre-installation tests

See Guidance on compiling Schedules 6 (Quality assurance and validation) and 13 (Pre-installation tests and procedures)

22.1 The **Project Manager** shall be entitled at all times to have access to any place where work under the **Contract** is being carried out or **Materials** are being manufactured or fabricated for the purpose of checking the progress of manufacture or fabrication, inspecting and observing the work or carrying out of tests on the work or **Materials**. The **Project Manager** shall give reasonable notice of the access and facilities he requires and the **Contractor** shall secure such access and facilities both at those premises under his own control and those under the control of his **Subcontractors**.

22.2 The **Contractor** shall arrange and be responsible for all factory and other off-site tests and those, if any, carried out at the **Site** prior to installation, including those listed in Schedule 13, and shall provide the **Project Manager** with copies of the test results in accordance with the quality assurance system described in Schedule 6.

22.3 Whenever the **Contractor** is ready to conduct a pre-installation test he shall notify the **Project Manager** of the place and time at which he intends to conduct it, the time being not less than ten days later than the date of such notice. The **Contractor** shall conduct the test at the time and place so notified.

If the **Project Manager** or his nominee does not attend the test, the **Contractor** shall nevertheless be entitled to conduct it in his absence and copies of the test results provided in accordance with the quality assurance system described in Schedule 6 shall be deemed to be a correct record of the test.

22.4 The **Project Manager** may require any additional pre-installation test not described in Schedule 13 to be conducted for the purpose of establishing whether any work or **Materials** are in accordance with the **Contract**. The **Project Manager** shall notify the **Contractor** of any such requirement by means of a **Variation Order** in sufficient time to enable the **Contractor** to conduct such test without impeding the performance of his other obligations under the **Contract**. The provisions of Sub-clauses 22.2 and 22.3 shall apply to any such test.

22.5 The **Contractor** shall provide or shall arrange for the provision of all labour, materials and equipment necessary for the proper carrying out of all pre-installation tests under this Clause.

22.6 If any item of work or **Materials** should fail to pass any pre-installation test, the **Contractor** shall either rectify or replace the relevant item and, unless the **Project Manager** dispenses with a repetition of the test, shall repeat the test following a further notice given under Sub-clause 22.3. The length of such notice may however be less than ten days if a lesser period is reasonable in the circumstances.

22.7 Unless the repeat test is undertaken by a **Subcontractor** without further cost to the **Contractor** or the cost of the repeat test is paid by the **Subcontractor** to the **Contractor**, the **Contractor's** cost of any repeat test shall form part of the **Contract Price**.

22.8 The rights of the **Project Manager** under this Clause may be exercised by any duly authorised nominee of the **Project Manager**.

23. The Site

See Guidance on compiling Schedule 1 (Description of the Works)

23.1 The **Purchaser** shall give the **Contractor** access to and possession of the **Site** or parts of the **Site**:

(a) by or before the date or dates specified therefor in the **Contract**; or

(b) if no such date or dates are specified, then in accordance with the **Approved Programme**; or

(c) if there be no such **Approved Programme**, then in reasonable time to permit the **Contractor** to perform his obligations under the **Contract**.

23.2 Unless the **Contract** otherwise provides, the **Purchaser** shall procure that the **Site** includes a suitable right of access for the **Contractor** from a convenient point on the public highway.

The **Contractor** shall satisfy himself as to the suitability and availability of routes to the **Site** and the power, water and other utilities which he chooses to use. The **Contractor** shall obtain any permission that may be required from the relevant authorities for the use of such route or utility. The **Purchaser** does not guarantee and makes no representation as to the suitability or availability of any route or utility.

23.3 The **Contractor** shall use every reasonable means to prevent any of the roads (including temporary roads), railways (if any), and all associated bridges and other structures connecting with or on routes to the **Site** from being damaged by any traffic generated by the **Contractor** or any **Subcontractor**. In order to avoid unnecessary damage to any such roads, railways, bridges or other structures, the **Contractor** shall select appropriate routes and vehicles or restrict and distribute loads such that the movement to or from the **Site** of **Contractor's Equipment**, **Materials** or any fabricated item, module or structure intended for incorporation into the **Plant** is limited as far as reasonably possible. The **Contractor** shall indemnify the **Purchaser** against all claims for damage to any such roads, railways, bridges or other structures caused by such traffic, including, but not limited to, claims made by any competent authority directly against the **Purchaser** pursuant to any applicable **Legislation**.

If the **Works** involve the use by the **Contractor** of waterborne transport, this Sub-clause shall be construed as though 'road' or 'route' includes a river, canal, lock, dock, sea wall or other structure related to a waterway, berth or port and 'traffic' includes any kind of waterborne craft, and shall have effect accordingly.

23.4 The **Contractor** shall ensure that every aspect of, and all operations connected with, the **Works** are carried out so as not to damage, or to interfere unnecessarily with, any of the operations being carried out by others on or about the **Site**, or with the convenience of the public or the access to, or to the occupation and use of, any public or private property including, but not limited to, public and private rights of way.

23.5 The **Contractor** shall permit the **Purchaser** and the **Project Manager** and their employees and agents to enter the **Site** at all reasonable hours for the purpose of inspecting the **Works**, receiving training, or performing their functions under the **Contract**. The **Contractor** shall also permit other contractors or suppliers engaged by the **Purchaser** to enter the **Site** and carry out work or deliver goods at all reasonable hours, but unless the **Contract** expressly provides otherwise, the **Contractor** shall not be required to allow them to have such access or facilities as would unreasonably impede his performance of the **Contract**.

23.6 Subject to Sub-clause 23.5, the **Contractor** shall not permit to enter the **Site** other persons whose presence on the **Site** is not necessary for the carrying out of the **Works** nor for the carrying out of other work on behalf of the **Purchaser** without the prior consent of the **Purchaser** or the **Project Manager**.

23.7 The **Contractor** shall take full account of and abide by the **Purchaser's** site and security rules for contractors and ensure that his **Subcontractors** do likewise.

24. Delivery to Site

24.1 The **Contractor** shall be responsible for the delivery to the **Site** of all **Materials** and **Contractor's Equipment** other than materials (if any) to be provided by the **Purchaser**, for which the delivery arrangements shall be as set out in Schedule 3 (Responsibilities of Purchaser).

24.2 Unless otherwise agreed with the **Project Manager**, the **Contractor** shall not deliver or cause to be delivered anything to the **Site** until he has provided adequate facilities at the **Site** for the proper unloading and storage thereof.

24.3 If the **Contractor** wishes any **Materials** or **Contractor's Equipment** to be delivered to the **Site** before the time, if any, specified in the **Approved Programme**, he shall obtain the consent of the **Project Manager** before such delivery is made.

24.4 If any **Materials** or **Contractor's Equipment** are to be delivered to the **Site** seven days or more after the time, if any, specified in the **Approved Programme**, the **Contractor** shall draw this to the attention of the **Project Manager** in writing immediately upon his becoming aware of the delay.

25. Ownership of Materials

See Guide Note J (Ownership)

25.1 The ownership of **Materials** shall pass to the **Purchaser** at whichever is the earlier of the following:

(a) upon delivery to the **Site**; or

(b) when the **Contractor** becomes entitled to any payment in respect of such **Materials** and such payment has been made.

The **Contractor** shall insert such provisions in subcontracts as will cause this to happen.

25.2 Whenever the ownership of any **Materials** passes to the **Purchaser** prior to delivery to the **Site**, the **Contractor** shall arrange for the **Materials** to be marked as the **Purchaser's** property and ensure that they are stored and handled separately from other materials.

25.3 When so requested by the **Project Manager**, the **Contractor** shall provide proof of title and its transfer to the **Purchaser**.

26. Health, Safety and Environment

See Guidance on compiling Schedules 4 (Health and Safety) and 5 (Environmental protection and waste disposal)

26.1 The **Purchaser** and the **Contractor** shall comply with all health, safety and environmental **Legislation** issued by any duly appointed authority having jurisdiction over the **Site** and/or the **Works**. Without prejudice to the generality of the foregoing obligation, the **Works** shall be carried out in compliance with the Health and Safety at Work Act 1974 (as amended by Schedule 2 to the Consumer Protection Act 1987) and all applicable Regulations made thereunder, including without limitation the Construction (Design and Management) Regulations 1994 or any statutory re-enactment or amendment thereof for the time being in force.

26.2 The **Contractor** shall be responsible for the safety of his own operations and those of any **Subcontractor** and shall ensure that:

(a) a safe working environment is maintained;

(b) all persons in proximity to the **Site**, whether or not employed by the **Contractor** or any **Subcontractor**, are properly protected from risk of injury and danger to health arising out of or in connection with the carrying out of the **Works**;

(c) all property under his control is properly protected from damage or loss;

(d) any hazardous material for which he is responsible is safely contained or removed from the **Site**.

26.3 The **Contractor** shall ensure that his employees, **Subcontractors**, and anybody working under their control shall be conversant with, and shall at all times comply strictly with, the requirements set forth in Schedules 4 and 5 and any site safety regulations, safe working procedures and health and safety instructions issued to the **Contractor** from time to time by the **Project Manager** or the **Purchaser**.

26.4 Without prejudice to the **Contractor's** general responsibility for safety as set forth in Sub-clause 26.1, the **Contractor** shall:

(a) take due notice of all instructions or advice given by the **Purchaser's** designated Health and Safety Supervisor;

(b) ensure that no hazardous, unsafe, unhealthy or environmentally unsound condition or activity over which he has control occurs at the **Site**; and

(c) nominate a competent person to be responsible for the proper observance of all safety requirements and procedures during work on the **Site**.

26.5 The **Contractor** shall be liable for and indemnify the **Purchaser** against all claims, damages and costs associated with any environmental pollution or hazardous material arising from the **Contractor's** activities at the **Site**, including the cost of cleaning up and remedying the effects of such pollution or escape of hazardous material, except when such pollution or escape:

(a) occurs as the direct result of compliance with any **Decision** of the **Purchaser** or the **Project Manager**; or

(b) is existing at the **Site** and the cleaning up and remedying the effect thereof could not reasonably have been foreseen by an experienced contractor possessed of all the information which the **Contractor** had or could have obtained by visual inspection of the **Site** or by reasonable enquiry.

27. Site services

See Guide Note G (Site services and working conditions)

27.1 The **Contractor** shall provide all **Materials**, **Contractor's Equipment**, labour, consumables, services and facilities that may be necessary at the **Site** for the proper carrying out and completion of the **Works** other than those, if any, which the **Contract** requires the **Purchaser** to provide.

27.2 Without prejudice to the above Sub-clause, the facilities and services to be provided by the **Contractor** shall include but not be limited to: temporary roads and parking; temporary offices, stores and warehousing; communications; sanitary and canteen facilities for persons engaged in the **Works**; cleaning and other janitorial services; and all necessary or specified safety, security, fire-fighting, first aid and other medical facilities.

27.3 All temporary buildings, structures, equipment and facilities provided by the **Contractor** at the **Site** shall be properly maintained and cleaned by the **Contractor** and he shall at all times during the carrying out of the **Works** keep the **Site** in a safe, clean and orderly condition and reasonably free from waste materials and rubbish.

27.4 If the **Contract** states that the **Purchaser** shall provide any services or facilities for use by the **Contractor**, then unless otherwise stated, the **Contractor** shall have the use thereof free of charge. Such use by the **Contractor** shall be subject to any condition that the **Project Manager** may reasonably require.

27.5 Any certificate or other document relating to **Contractor's Equipment** and facilities which the **Contractor** is by law required to have, make or obtain shall, if required, be produced for inspection by the **Project Manager** or **Purchaser**.

27.6 The **Contractor** shall not remove **Contractor's Equipment** from the **Site** without the prior consent of the **Project Manager**, but the **Project Manager** shall not refuse consent if the **Contractor's Equipment** is no longer required for the carrying out of the **Works**. If in the **Project Manager's** opinion it is reasonable to do so he may instruct the **Contractor** to remove the **Contractor's Equipment** from the **Site** subject to a reasonable time to comply with such instruction. Thereafter, the cost of such **Contractor's Equipment** shall not be charged to the **Purchaser**.

28. Site working conditions

See Guide Note G (Site services and working conditions)

28.1 Subject to the provisions of Sub-clause 28.3, the **Contractor** shall pay rates of wages and allowances and observe hours and conditions of employment which are not less favourable than those established for the relevant trade or industry by any general agreement or award applying to the district where any work for the **Contract** is being carried out and shall impose the like obligations on his **Subcontractors**.

28.2 The **Contractor** shall at all times keep the **Project Manager** informed of any matter likely to affect industrial relations at the **Site**.

28.3 The **Contractor** shall comply with the terms of any general agreement made between representatives of workmen and employers who will be engaged in work at the **Site** as to the rates of wages and allowances to be paid to and general working conditions for workmen employed at the **Purchaser's** premises where the **Site** is situated, provided that such agreement was either already in force at the date of tender as stated in the Agreement or is one to which the **Contractor** is already or becomes a party.

28.4 The **Contractor** shall forthwith secure the removal from the **Site** of any person (other than his **Site Manager** or a member of his **Site** staff, to whom the provisions of Sub-clause 12.5 apply) whom the **Project Manager** shall by notice require to be removed upon the grounds of misconduct, incompetence or serious breach of any of the requirements referred to in Sub-clauses 23.7 and 26.3. Any such notice given by the **Project Manager** shall be given in good faith and shall be final, binding and conclusive and not capable of being resolved under Clause 45 (Disputes) or reviewed, revised or reversed by an Adjudicator under Clause 46 (Adjudication) or by reference to an **Expert** under Clause 47 (Reference to an Expert), or by an Arbitrator under Clause 48 (Arbitration) or in court proceedings. The **Project Manager** shall not delegate such power to any **Project Manager's Representative**. The **Purchaser** shall have no obligation to reimburse the **Contractor** for the cost of replacing such person.

28.5 Subject to anything to the contrary stated in the **Contract**, the **Contractor** shall not, except in an emergency, carry out any work on the **Site** at times outside the normal working hours, if any, specified in Schedule 1 (Description of the Works) without the prior agreement of the **Project Manager**.

29. Meetings

29.1 Meetings to review progress and discuss matters relating to the **Works** shall be held on a regular basis on dates to be fixed by the **Project Manager**. The **Contractor** also may request that a meeting be held at another time if the circumstances require it, and the **Project Manager** shall not unreasonably reject such a request. Meetings shall be attended by the **Project Manager** or the **Project Manager's Representative** and by the **Contract Manager** or **Site Manager** and may be attended by other persons, including representatives of **Subcontractors**, as appropriate.

29.2 The **Project Manager** shall provide a minute secretary for each meeting. Within seven days of the meeting the **Project Manager** shall give three copies of the minutes signed by him to the **Contract Manager**. If the **Contract Manager** accepts the minutes as a sufficient and accurate record, he shall sign one copy and return it to the **Project Manager** within a further seven days. If not, the **Contract Manager** shall agree any modifications with the **Project Manager** and the amended copies shall be signed by both of them. If agreement cannot be reached upon any modification to the minutes, the disagreements remaining shall be noted in writing and attached to all copies of the minutes.

29.3 Only minutes signed by both the **Contract Manager** and the **Project Manager** shall constitute approved minutes for the purposes of Sub-clause 5.3.

30. Care of the Works

See Guide Notes K (Insurance) and P (Limitations of Contractor's liability)

30.1 Unless otherwise agreed, the **Plant** and **Materials** shall be under the care, direction and control of the **Contractor** until the **Plant** (or any specified section thereof) is taken over by the **Purchaser**. Access to the **Site** and all activities on it pursuant to the **Contract** (other than the activities of the **Purchaser** or the **Purchaser's** agents) shall, subject to the terms of Sub-clause 23.1, be under the direction and control of the **Contractor** until the whole of the **Plant** is taken over by the **Purchaser**.

30.2 The **Contractor** shall make good any loss or damage that may occur to the **Materials**, **Plant** or **Documentation**:

(a) before the **Plant** (or each specified section thereof) is taken over by the **Purchaser**, howsoever caused; and

(b) after the **Plant** (or each specified section thereof) has been taken over by the **Purchaser** but before the issue of the **Final Certificate**, when the loss or damage results from any cause or operation described in Sub-clause 31.1 (a) or (b);

other than loss or damage arising from the **Purchaser's** risks as defined in Sub-clause 30.4.

Without prejudice to the **Contractor's** liability under Clause 37 (Liability for Defects), the **Contractor's** liability under this Sub-clause 30.2 for the cost of making good loss or damage shall be limited to the amount recoverable under the policy or policies of insurance effected by the **Purchaser** in accordance with Sub-clauses 31.1 and 31.2. However, except in the case of loss or damage arising from the **Purchaser's** risks as defined in Sub-clause 30.4, the **Contractor** shall bear the deductible stated in the Agreement for each and every claim.

30.3 To the extent that any loss or damage arises from any of the **Purchaser's** risks, the **Contractor** shall, if so instructed by the **Project Manager**, rectify the loss or damage and the cost thereof shall form part of the **Contract Price**.

30.4 The '**Purchaser's** risks' under this Clause are loss or damage due to:

(a) the use or occupation of the **Plant** by the **Purchaser**, his employees or agents, or other contractors (not being employed by the **Contractor**);

(b) any design or information provided by the **Purchaser** (other than in any design verified by the **Contractor** pursuant to his obligations under the **Contract**);

(c) any wrongful or negligent act or omission of the **Purchaser**, his consultants, employees or agents, or other contractors (not being employed by the **Contractor**);

(d) riot, war, invasion, act of foreign enemies, hostilities (whether or not war be declared), terrorism, civil unrest, civil war, rebellion, revolution, insurrection, military or usurped power, or similar events;

(e) ionising radiations or contamination by radioactivity from any nuclear fuel or from any nuclear waste, from the combustion of nuclear fuel, radioactive, toxic, explosive or other hazardous properties of any explosive nuclear assembly or nuclear component thereof; or

(f) pressure waves caused by aircraft or other aerial devices travelling at sonic or supersonic speed.

30.5 The **Contractor** shall assume entire responsibility for and shall indemnify the **Purchaser** against all losses, liabilities, claims and costs arising directly or indirectly out of or in connection with or as a result of, the death or bodily injury to any person in the employment of the **Contractor** or of any **Subcontractor** and loss or damage to the **Contractor's Equipment** or other property of the **Contractor** or any **Subcontractor**.

30.6 The **Purchaser** shall assume entire responsibility for and shall indemnify the **Contractor** and all **Subcontractors** against all losses, liabilities, claims and costs arising directly or indirectly out of or in connection with or as a result of death or bodily injury to any person in the employment of the **Purchaser** or the **Project Manager**.

30.7 Subject to the provisions of Sub-clauses 30.2, 30.3 and 30.4, the **Contractor** shall be liable for and shall indemnify the **Purchaser** against any loss of or damage to the property of the **Purchaser** and his **Affiliates** (other than **Materials, Plant** or **Documentation**) from any cause arising out of the performance of the **Works** up to a limit of such amount as may be stated in the Agreement or, if no such sum is stated, up to £5,000,000 (five million pounds) in respect of any one incident or series of incidents arising from any one event, and the **Purchaser** shall indemnify the **Contractor** and all **Subcontractors** from any sums in excess of such amount.

30.8 The **Contractor** shall be liable for and shall indemnify the **Purchaser** against any loss or damage to property of third parties and death of or injury to third parties to the extent that such loss, damage, death or injury results from any wrongful or negligent act or omission of the **Contractor**, his employees, agents or consultants.

30.9 The **Purchaser** shall indemnify the **Contractor** and all **Subcontractors** in respect of any legal liability for loss of or damage to property of the **Purchaser** (other than that to the **Plant** or **Materials** prior to their being taken over by the **Purchaser**) caused by fire, explosion and all other perils insured by the **Purchaser** as referred to in Sub-clause 31.2 to the extent that such liability exceeds for any one incident or series of incidents arising from one event the amount recoverable under such insurance.

31. Insurance

See Guide Note K (Insurance)

31.1 The **Purchaser** shall effect and maintain a policy or policies of insurance in the joint names of the **Purchaser**, **Project Manager**, **Contractor** and **Subcontractors**, upon the **Materials**, **Plant** and **Documentation** to the full cost of their replacement (or such other sum as may have been agreed) for any loss or damage thereto resulting from any one incident or series of incidents arising from any one event until the **Plant** or any specified section thereof has been taken over by the **Purchaser** under the provisions of Clause 33 (Taking over), and thereafter until the issue of the **Final Certificate** for any loss or damage resulting from:

(a) any cause occurring prior to taking over;

(b) any testing or other operations after taking over carried out or supervised by the **Contractor** or any **Subcontractor** for the purpose of complying with their obligations under the **Contract**.

31.2 The policy or policies shall cover all normally insurable risks including loss or damage during transport by land, sea or air, but excluding:

(a) the cost of making good defective designs or specifications and faulty workmanship or **Materials** (provided that this exclusion shall not apply to loss or damage resulting therefrom); and

(b) those risks specified in Sub-clauses 30.4 (c), (d), (e) and (f), and such other common exclusions as may be accepted by the **Purchaser**.

The deductible under the policy or policies shall be as stated in the **Agreement**.

Except as may be agreed otherwise, the **Purchaser**, as policyholder shall represent the insured parties in all matters relating to the policy or policies. The **Purchaser** undertakes not to give any release or compromise any claim affecting the interests of the **Contractor** without the prior written consent of the **Contractor**.

31.3 The **Purchaser** shall insure all of his property and that of his **Affiliates** on or adjacent to the **Site** (other than the **Plant** and **Materials** prior to taking over) against all normal risks including fire, lightning, explosion, storm, tempest, flood, earthquake, aircraft (or articles dropped therefrom), terrorism, riot and civil commotion, and the interest of the **Contractor** shall be noted thereon.

31.4 The **Contractor** shall effect and maintain a policy or policies of insurance covering:

(a) his liability to the **Purchaser** and his **Affiliates** under Sub-clause 30.7;

(b) loss or damage to **Contractor's Equipment**; and

(c) his legal liability to any third party for such sum as the **Contractor** considers appropriate, but in any event not less than £5,000,000 (five million pounds).

The above insurance policy or policies shall contain an indemnity to principals clause and shall become effective upon commencement of the **Works** and shall remain in place until the issue of the last **Final Certificate**.

31.5 Before the **Contractor** starts to work on the **Site** each party shall provide the other details of his insurances as provided herein. Each party shall also provide details in a timely manner of any additions or restrictions thereto which may be made from time to time. Each party shall provide to the other sufficient evidence of the payment of premiums.

32. Completion of construction

See Guidance on compiling Schedule 14 (Criteria for the completion of construction) and Appendix A (Sample Construction Completion Certificate)

32.1 If the **Contract** provides for the completion of construction of the **Plant** to be by specified sections, the provisions of Sub-clauses 32.2 to 32.5 shall apply as if the reference therein to **Plant** were a reference to a specified section.

32.2 As soon as the **Plant**, or any part thereof, is in the opinion of the **Contractor** substantially complete and ready for inspection, he shall so notify the **Project Manager** by means of a draft Construction Completion Certificate, listing the parts of the **Plant** considered to be substantially complete, and also the criteria listed in Schedule 14 which apply to such parts of the **Plant**. The **Contractor** shall propose a programme for such inspection and for any tests, commencing not sooner than seven days nor later than fourteen days after the date of the notice unless the **Project Manager** agrees otherwise.

32.3 Upon satisfactory completion of any such inspection and tests, the **Project Manager** shall complete and issue to the **Purchaser** and to the **Contractor** copies of the completed Construction Completion Certificate confirming that the **Contractor** has demonstrated to the **Project Manager** that the **Plant** or part thereof is substantially complete and in a condition such that any procedures needed to be carried out before the **Plant** is put into operation may be safely carried out. Such Construction Completion Certificate may include a note of any minor items requiring completion before the issue of a **Take-Over Certificate**, as provided in Clause 33 (Taking over).

32.4 If the **Project Manager** is not satisfied that the **Plant** or part thereof referred to in Sub-clause 32.2 is substantially complete (including passing any test required by Schedule 14), he may endorse the draft Construction Completion Certificate accordingly, stating in what way the **Plant** or part thereof is not in accordance with Schedule 14. The **Contractor** shall then complete the **Plant** or part thereof as necessary and shall repeat the procedure described in Sub-clause 32.2.

32.5 The cost of any repeat inspection or test shall form part of the **Contract Price**, except where:

(a) the repeat inspection or test is undertaken by a **Subcontractor** without further cost to the **Contractor** or the cost of the repeat test is paid by the **Subcontractor** to the **Contractor**; or

(b) the repeat inspection or test is due to the **Contractor's** negligence or the failure of the **Contractor**, his employees or agents to exercise the requisite degree of skill and care which would ordinarily be exercised by properly qualified and reasonably competent persons engaged in the same or a similar business of the **Contractor**.

32.6 If in Schedule 11 (Times of completion) a date is specified for the completion of construction of the **Plant** as a whole, the **Project Manager** shall, as soon as he has signed all the Construction Completion Certificates in accordance with Sub-clause 32.3, issue a Construction Completion Certificate for the **Plant** as a whole.

32.7 If the **Contractor** considers that the **Project Manager's Decision** concerning the issue of a Construction Completion Certificate is unreasonable and if the **Contractor** and the **Project Manager** cannot agree on the matter, the dispute may be referred to an **Expert** in accordance with Clause 47 (Reference to an Expert).

33. Taking over

See Guide Note S (Completion, taking over, testing and start-up), Guidance on compiling Schedule 15 (Take-over procedures) and Appendix B (Sample Take-Over Certificate)

33.1 If the **Contract** provides for the **Plant** to be taken over by specified sections, the provisions of this Clause shall apply as if the references therein to the **Plant** were references to a specified section.

33.2 As soon as the construction of any part of the **Plant** has been demonstrated to be complete in accordance with the provisions of Sub-clause 32.3 and is, in the opinion of the **Contractor**, ready for the conduct of any of the take-over procedures, which may include take-over tests, specified in Schedule 15, the **Contractor** shall so notify the **Project Manager** and shall specify a time not sooner than seven days and not later than fourteen days after the date of the notice when the **Contractor** intends to begin to conduct such procedures. If the **Project Manager** requires the **Contractor** to carry out any take-over procedures which are not included in Schedule 15, such requirements shall be treated as a **Variation**.

33.3 Unless otherwise agreed between the **Project Manager** and the **Contractor**, the **Contractor** shall begin such procedures at the time notified and conduct them in accordance with the quality assurance system, if any, described in Schedule 6 (Quality assurance and validation). The **Project Manager** shall be entitled to attend and observe them, and the **Contractor** shall give the **Project Manager** every reasonable facility to satisfy himself as to the results of any take-over tests.

33.4 Unless the **Contract** otherwise provides, or the **Contractor** and the **Project Manager** otherwise agree, the **Contractor** shall provide all labour, materials and equipment necessary for the proper conduct of the take-over procedures.

33.5 If any part of the **Plant** fails to pass a take-over test, the **Contractor**, after making such adjustments as he considers necessary, shall repeat such test in the presence of the **Project Manager** at a time agreed between the **Contractor** and the **Project Manager**, or notified in accordance with Sub-clause 33.2. If the **Project Manager** is of the opinion that any such adjustments made by the **Contractor** make it desirable, he may require the **Contractor** to repeat any take-over test which has already been successfully carried out, and the **Contractor** shall do so. Save as aforesaid the **Contractor** shall not be bound to repeat for the purposes of this Clause any take-over test that has already been successfully carried out.

The provisions of Sub-clause 32.5 shall apply to any repetition of any take-over test.

33.6 If the **Project Manager** does not attend any take-over procedure in accordance with Sub-clause 33.3, the **Contractor** shall nevertheless be entitled to conduct it in his absence, and copies of the test results provided in accordance with the quality assurance system, if any, described in Schedule 6 shall be deemed to be a correct record of the test.

33.7 As soon as any minor items referred to in Sub-clause 32.3 have been completed and all the procedures specified in Schedule 15 have been successfully carried out subject to the provisions of Sub-clause 33.10, including any which affect the operability or safety of the **Plant**, the **Project Manager** shall issue to the **Contractor** with a copy to the **Purchaser** a **Take-Over Certificate** for the **Plant** stating that the **Contractor** has satisfied the requirements of the **Specification** and Schedule 15, whereupon the **Plant**, apart from any parts that are excluded from the taking over by the terms of the Certificate, shall be at the risk of the **Purchaser**. The **Purchaser** shall thereupon be responsible for the care, safety, operation, servicing and maintenance of the **Plant** so certified and if Clause 35 (Performance tests) applies, shall start it up and prepare for and carry out the performance tests.

The **Take-Over Certificate** may include a list of minor items still to be completed by the **Contractor**, and a list of take-over procedures omitted by operation of Sub-clause 33.10.

If the **Contractor** considers that the **Project Manager's Decision** concerning the issue of any **Take-Over Certificate** was unreasonable and if the **Contractor** and the **Project Manager** cannot agree on the matter, the dispute may be referred to an **Expert** in accordance with Clause 47 (Reference to an Expert).

33.8 As soon as shall be reasonably possible after the issue of the **Take-Over Certificate** the **Contractor** shall complete to the satisfaction of the **Project Manager** all items noted on the **Take-Over Certificate** as not being complete at the date of the said Certificate. If the **Contractor** shall fail so to do the **Project Manager** may arrange for such work to be carried out by others in accordance with Clause 37 (Liability for Defects).

33.9 Notwithstanding the failure of the **Plant** to pass any take-over test or if certain procedures have not been carried out, the **Project Manager** may with the **Contractor's** consent issue a **Take-Over Certificate** in respect of the **Plant**. Such a Certificate may either exclude from taking over such parts of the **Plant** as are specified therein, being parts for which the specified procedures have not been carried out, or may specify the take-over tests which the **Plant** has failed to pass. In either case the **Contractor** shall remain liable until the end of the **Defects Liability Period**, should the **Project Manager** so require, to undertake any omitted procedures or to repeat any tests which the **Plant** has not passed and the **Contractor** shall become entitled to a separate **Take-Over Certificate** as soon as the applicable take-over procedures have been completed satisfactorily.

33.10 If in the opinion of the **Contractor** he has been prevented from carrying out any take-over procedure as provided in this Clause by reason of any act or omission of the **Purchaser** or of the **Project Manager**, or of some other contractor employed by the **Purchaser**, then when any other relevant take-over procedures have been successfully completed, the **Contractor** may request the **Project Manager** in writing to issue a **Take-Over Certificate**. The **Project Manager** shall, if in his opinion such action is reasonable, issue such a Certificate which shall have effect as set out in Sub-clause 33.7. The **Contractor** shall during the **Defects Liability Period** carry out any omitted take-over procedure as and when required by the **Project Manager** who shall give fourteen days notice of such requirement. If in the opinion of the **Contractor** the **Project Manager's** refusal to issue such a **Take-Over Certificate** is not reasonable, and the **Contractor** and the **Project Manager** cannot agree on the matter, the dispute may be referred to an **Expert** in accordance with Clause 47 (Reference to an Expert).

33.11 Any effect on the results of a deferred take-over procedure conducted in accordance with Sub-clauses 33.9 or 33.10 which can reasonably be shown to be due to the prior use of the **Plant** by the **Purchaser** shall be taken into account in assessing such results.

33.12 If the **Purchaser** has taken over the **Plant** as a result of the issue of a **Take-Over Certificate** in accordance with the provisions of Sub-clauses 33.9 or 33.10 and subsequently the **Plant** fails to satisfy the requirements of any relevant take-over procedure which had previously failed or been omitted then, subject to the **Contractor** having been given a reasonable opportunity to carry out such take-over procedure including making any necessary adjustments to the **Plant**, the **Project Manager** may issue a revocation certificate to the **Contractor** with a copy to the **Purchaser** stating that any **Acceptance Certificate** which has been deemed to be issued in terms of Sub-clause 36.1 is thereby revoked. Thereupon the rights and obligations of the **Contractor** and **Purchaser** shall revert as far as is possible to those that existed before the issue of such **Acceptance Certificate**.

33.13 After the issue of a **Take-Over Certificate**, and until the **Final Certificate** has been issued, the **Contractor** shall have the right of access during all reasonable working hours at his own risk and cost, by himself or his duly authorised representatives whose names shall previously have been communicated in writing to the **Project Manager**, to inspect any part of the **Plant** and to take notes of the related working, performance and maintenance records. Subject to the **Project Manager's** approval, the **Contractor** may at his own risk and cost make any tests which he considers desirable.

34. Site clearance

See Guidance on compiling Schedule 5 (Environmental protection and waste disposal)

34.1 Unless otherwise agreed with the **Project Manager**, the **Contractor** shall within fourteen days after the **Purchaser** has taken over the **Plant** remove from the **Site** all waste and debris, **Contractor's Equipment**, including temporary buildings and structures, and other things of his, other than those that are necessary for the proper performance of the **Contractor's** obligations and shall leave the **Site** in an orderly and safe condition. Where the **Plant** is taken over by specified sections, the **Contractor** shall remove such things from within the defined perimeter of each such section as it is taken over, or if there be no such perimeter, then from the area about the section as prescribed by the **Project Manager**.

34.2 As soon as the whole of the **Plant** has been **Accepted**, the **Contractor** shall remove all his things from the **Site** other than such things, if any, as are then known to be required for making good defects under Clause 37 (Liability for Defects) and any **Materials** which are subject to Sub-clause 34.5.

34.3 No **Materials** shall be removed from the **Site** unless:

(a) they are not in accordance with the **Contract** and are being removed for the purpose of:

(i) modification; or

(ii) replacement;

or

(b) the **Project Manager** has previously consented to their removal; or

(c) in the case of bulk materials, they are surplus to the requirements of the **Works**.

34.4 The property in all **Materials** removed from the **Site** in accordance with Sub-clauses 34.3 (a)(ii), (b) and (c) shall revert to the **Contractor** upon removal. **Materials** falling within the description of Sub-clause 34.3 (a)(i) shall be marked as the **Purchaser's** property and stored and labelled separately from other materials.

34.5 If the **Purchaser** shall so require, the **Contractor** shall remove surplus materials and dispose of them on behalf on the **Purchaser**. Any surplus **Materials** that have been held in care by the **Contractor** and which are to remain the property of the **Purchaser** shall be handed over by the **Contractor** before he leaves the **Site**.

35. Performance tests

See Guide Note S (Completion, taking over, testing and start-up), Guidance on compiling Schedules 16 (Performance tests and procedures) and 17 (Performance guarantees and damages for failure)

35.1 If under the **Contract** the **Contractor** makes specific guarantees in respect of the performance of the **Plant**, verifiable by performance tests, the provisions of this Clause shall apply. Otherwise this Clause shall not form part of the **Contract**.

35.2 If the **Contract** provides for the performance of the **Plant** to be tested in specified sections, the provisions of this Clause shall apply as if the references therein to the **Plant** were references to a specified section.

35.3 The performance tests to be carried out on the **Plant** shall be those specified in Schedule 16. If any unspecified test is subsequently proposed, the **Contractor** and the **Project Manager** shall discuss whether such tests should be carried out. If they agree that the test should be carried out, then the test should be treated as a **Variation** in accordance with Clause 16 (Variations) with the exception of Sub-clause 16.7.

35.4 The performance tests shall be carried out by the **Purchaser** in the presence of the **Contractor** as soon as is practicable after he has taken over the **Plant**, using suitably trained and experienced employees and in accordance with the manuals provided by the **Contractor** pursuant to Clause 21 (Documentation) and such other instruction as the **Contractor** may give in the course of carrying out such tests. If any such instruction conflicts in any way with the manuals, the **Contractor** shall issue it in writing in the form of an amendment. The performance tests shall be carried out as far as practicable under the conditions, if any, detailed in the **Specification**.

35.5 The **Purchaser** shall give the **Contractor** at least fourteen days notice of his readiness to carry out the performance tests, including a proposal for the time at which the tests should commence. The **Contractor** shall confirm, at least seven days before the time proposed by the **Purchaser**, his readiness to attend the tests.

35.6 Every performance test shall be carried out to completion (as may be specified in Schedule 16) unless an authorised representative of the **Purchaser**, the **Project Manager** or the **Contractor** shall order it to be stopped because its continuance would be unsafe or unacceptable to either party.

35.7 If the **Plant** fails to pass any performance test or if any performance test is stopped before its completion such test shall, subject to Sub-clause 35.9, be repeated as soon as practicable thereafter. Meanwhile the **Purchaser** shall have the right to operate the **Plant**. The **Purchaser** shall permit the **Contractor** to make adjustments and modifications to any part of the **Plant** before the repetition of any performance test and shall, if the **Contractor** reasonably requires, shut down any part of the **Plant** for such purpose and restart it after completion of the adjustments and modifications, which shall be made by the **Contractor** with all reasonable speed. The timing of such shutdown shall be agreed between the **Contractor** and the **Project Manager**.

The **Contractor** shall, if so required by the **Project Manager**, submit to the **Project Manager** for his approval details of any adjustment or modification which he proposes to make.

All costs arising from or in connection with such adjustment or modification shall form part of the **Contract Price** except where:

(a) the adjustment or modification is made by a **Subcontractor** without further cost to the **Contractor** or the cost of such work is paid by the **Subcontractor** to the **Contractor**; or

(b) the adjustment or modification was necessitated by a failure of the **Contractor** to exercise the requisite degree of skill and care which would ordinarily be exercised by properly qualified and reasonably competent persons engaged in the same or a similar business to that of the **Contractor**.

35.8 The results of the performance tests shall be compiled and evaluated jointly by the **Purchaser** or the **Project Manager** and by the **Contractor**. Feed and product rates and the consumption of utilities shall be averaged over the specified period for the relevant test, unless Schedule 16 otherwise provides. Any necessary adjustments to take account of actual operating conditions and measuring tolerances shall be made in accordance with the provisions of the Schedule. Any effect on the results of such performance tests which can reasonably be shown to be due to the prior use of the **Plant** by the **Purchaser** shall be taken into account in assessing such results.

In the event that the **Purchaser** and the **Contractor** are unable to agree on whether the **Plant** has passed the relevant performance tests, the dispute may be referred to an **Expert** in accordance with Clause 47 (Reference to an Expert).

35.9 If the **Plant** does not pass all relevant performance tests within the period stated in Schedule 16 from the date of the relevant **Take-Over Certificate**, and providing no unreasonable delay has been caused by the **Purchaser** in commencing or carrying out his obligations under Sub-clauses 35.4, 35.5 and 35.7 above, the **Purchaser** shall thereafter be free to operate the **Plant** as he sees fit, and the **Contractor** shall pay any liquidated damages due to the **Purchaser** in accordance with Schedule 17. Upon payment or allowance of such sum the **Contractor** shall become entitled to the issue of an **Acceptance Certificate** in respect of the **Plant** stating that the applicable liquidated damages have been paid in respect of the shortfall in performance.

35.10 If the **Plant** fails to pass any performance test and the **Contractor** in consequence proposes to make any adjustment or modification thereto, the **Project Manager** may notify the **Contractor** that the **Purchaser** wishes to defer such work until a time convenient to the **Purchaser,** in which case the period stated in Schedule 16 and referred to in Sub-clause 35.9 shall be extended by the length of time of the deferment. In such event the **Contractor** shall become entitled to the issue of an **Acceptance Certificate** as if the **Plant** had passed the performance test in question, but the **Contractor** shall remain liable to carry out the adjustment or modification and to satisfy the performance test within a reasonable time of being notified to do so by the **Project Manager**. Such **Acceptance Certificate** shall list the unsatisfied or uncompleted performance tests.

35.11 If the **Plant** has been prevented from passing its performance tests by matters beyond the **Contractor's** control, within the time limit stated in Schedule 16 and referred to in Sub-clause 35.9, the **Contractor** may apply to the **Project Manager** for the issue of an **Acceptance Certificate** in respect of the **Plant**. The **Project Manager** shall thereupon issue such a certificate both to the **Purchaser** and to the **Contractor** if:

(a) the **Project Manager** accepts the claim by the **Contractor** that but for reasons beyond the **Contractor's** control the **Plant** would have passed its performance tests; and

(b) the **Contractor**, if required by the **Purchaser**, has tendered a bond or guarantee to the **Purchaser** on the terms set out in Sub-clause 35.12.

The **Acceptance Certificate** shall list the unsatisfied or uncompleted performance tests.

35.12 The **Contractor** shall, if required by the **Purchaser** pursuant to Sub-clause 35.11, provide the **Purchaser** with a bond or guarantee in terms and from a bank or insurance company acceptable to the **Purchaser** guaranteeing:

(a) the repayment by the **Contractor** in accordance with Sub-clause 36.4 of any sum that may have become payable to him under the **Contract** upon the issue of the **Acceptance Certificate**; and

(b) the due payment by the **Contractor** of any sum that may become payable in the event of the **Plant** failing to pass the performance tests.

The cost of such bond or guarantee shall form part of the **Contract Price**.

35.13 All uncompleted work and all unsatisfied or uncompleted performance tests shall be carried out as soon as practicable after the issue of an **Acceptance Certificate** pursuant to Sub-clause 35.11. If matters beyond the **Contractor's** control continue to prevent the carrying out of the performance tests throughout the remainder of the relevant **Defects Liability Period**, then at the end of the said period such tests shall be deemed to have been satisfied unless it is otherwise established that the **Plant** was throughout the said period incapable of passing any such performance tests by reason of some default of the **Contractor**.

36. Acceptance

See Guide Note S (Completion, taking over, testing and start-up) and Appendix C (Sample Acceptance Certificate)

36.1 If under the **Contract**, the **Contractor** makes specific guarantees in respect of the performance of the **Plant**, verifiable by performance tests, the following Sub-clauses shall apply. Otherwise any **Take-Over Certificate** issued under the provisions of Clause 33 (Taking over) shall be deemed to be an **Acceptance Certificate** for the purposes of the **Contract** and Sub-clauses 36.2 to 36.6 shall not apply.

36.2 If the **Contract** provides for the performance of the **Plant** to be tested in specified sections, the provisions of this Clause shall apply as if the references therein to the **Plant** were references to a specified section.

36.3 Subject to the provisions of Sub-clauses 35.9, 35.10 and 35.11, as soon as the **Plant** has passed all the performance tests, the **Project Manager** shall issue to the **Contractor** a certificate (an '**Acceptance Certificate**') with a copy to the **Purchaser** stating that the **Plant** is accepted by the **Purchaser** as from the date thereof. The **Acceptance Certificate** shall list any known **Defects** which the **Contractor** is bound to make good under the provisions of Clause 37 (Liability for Defects) and any minor items still remaining to be completed following the issue of a **Take-Over Certificate** under Sub-clause 33.7.

36.4 Subject only to the provisions of Sub-clause 35.9, if the **Plant** fails to pass any performance test conducted pursuant to Sub-clause 35.13, the **Project Manager** may revoke the **Acceptance Certificate** by giving twenty-eight days notice to the **Purchaser** and the **Contractor**. If after the expiry of such notice the relevant performance tests have still not been passed or appropriate liquidated damages have not been paid, then any sum paid to the **Contractor** upon the **Acceptance** of the **Plant** shall be repaid by the **Contractor** to the **Purchaser** and the rights and obligations of the parties shall be as if the **Purchaser** had never accepted the **Plant**.

36.5 **Acceptance** of the **Plant** by the **Purchaser** shall not affect the obligations of the **Contractor** under Clause 37.

36.6 At any time after taking over the **Plant**, the **Purchaser** may elect to accept it by notice to the **Contractor** expressly stating that notice is given under this Sub-clause. The **Project Manager** shall not have authority to give such a notice on the **Purchaser's** behalf.

The **Project Manager** shall then issue an **Acceptance Certificate**. The **Acceptance** of the **Plant** under this Sub-clause shall constitute a waiver by the **Purchaser** of any remaining obligation of the **Contractor** to secure that the **Plant** passes the performance tests.

37. Liability for Defects

See Guide Note M (Liability for Defects)

37.1 If the **Contract** provides for the **Plant** to be taken over by specified sections, the provisions of this Clause shall apply as if the references therein to the **Plant** were references to a specified section.

37.2 Subject to the provisions of Sub-clause 37.12, if at any time before the **Plant** is taken over pursuant to Clause 33 (Taking over), or within a period of three hundred and sixty-five days after the date of the relevant **Take-Over Certificate** (the **Defects Liability Period**), the **Project Manager**:

(a) decides that any work done or **Materials** supplied by the **Contractor** or any **Subcontractor** is or are defective or not in accordance with the **Contract** (normal wear and tear excepted) or that the **Plant** or any part thereof is defective or incomplete or does not fulfil the requirements of the **Contract** (any such matter being herein called a '**Defect**'); and

(b) as soon as reasonably practicable gives to the **Contractor** notice in writing of the particulars of the alleged **Defects**;

the **Contractor** shall as soon as reasonably practicable make good the **Defects** so specified and the **Purchaser** shall so far as may be necessary place the **Plant** at the **Contractor's** disposal for this purpose. The **Contractor** shall, if so required by the **Project Manager**, submit his proposals for making good any **Defect** to the **Project Manager** for his approval.

37.3 The **Contractor's** cost of making good such a **Defect** shall form part of the **Contract Price**, except where:

(a) a **Defect** occurs in an item of the **Plant** supplied in accordance with Clause 9 (Assignment and subcontracting) and such **Defect** is made good by the **Subcontractor** without further cost to the **Contractor**, or the cost of making good such **Defect** is paid by the **Subcontractor** to the **Contractor**; or

(b) a **Defect** is due to the failure of the **Contractor** to exercise the requisite degree of skill and care which would ordinarily be exercised by properly qualified and reasonably competent persons engaged in the same or a similar business to that of the **Contractor**.

37.4 If a matter which the **Project Manager** has notified as a **Defect** is due to a failure of the **Purchaser** to operate or maintain the **Plant** in accordance with the operation and maintenance manuals provided by the **Contractor** and/or with good practice, the **Project Manager** shall treat the work to be done as a **Variation**.

37.5 If a **Defect** is made good after the issue of a **Take-Over Certificate**, the **Project Manager** may require the **Contractor** to repeat any appropriate take-over test for the purpose of establishing that the **Defect** has indeed been made good. The provisions of Sub-clause 32.5 shall apply to any such repeated test.

37.6 If in the course of making good any **Defect** which arises during the **Defects Liability Period** the **Contractor** repairs, replaces or renews any part of the **Plant**, the provisions of this Clause shall apply to the repair or to that part of the **Plant** so replaced or renewed and shall further apply until the expiry of a period of three hundred and sixty-five days from the date of such repair, replacement or renewal (the extended **Defects Liability Period**).

37.7 If the **Contractor** fails to make good within a reasonable time any **Defect** which he is liable to make good under Sub-clause 37.2, the **Purchaser** may, without prejudice to any other remedies or relief available to him under the **Contract**, proceed to do the work, provided that the **Purchaser** gives at least fourteen days notice of his intention in writing.

37.8 If the **Purchaser** reasonably requires that any **Defect** notified to the **Contractor** under Sub-clause 37.2 which arises during the **Defects Liability Period** be made good urgently and the **Contractor** is unable or refuses to comply within a reasonable time, then the **Purchaser** may, without prejudice to any other remedies or relief available to him under the **Contract**, proceed to do the work in such a manner as the **Project Manager** may decide, including the employment of a third party.

37.9 If the **Purchaser** has made good a **Defect** or completed a minor item of work in pursuance of either Sub-clauses 37.7 or 37.8 and if the cost of remedying the **Defect** is in accordance with Sub-clause 37.3 not to form part of the **Contract Price**, then the **Contractor** shall reimburse the **Purchaser** his reasonable cost of so doing provided that the **Purchaser** complies with the terms in Sub-clause 4.3. The **Project Manager** and the **Contractor** may agree the amount to be paid by the **Contractor**, or in the absence of agreement the **Project Manager** shall decide such amount as may be reasonable. Such amount shall be:

(a) deducted from any money that would otherwise be payable under the **Contract**; or

(b) due from the **Contractor** to the **Purchaser** as a debt.

If the **Contractor** cannot agree to such a **Decision** of the **Project Manager**, the dispute may be referred to an **Expert** in accordance with Clause 47 (Reference to an Expert).

37.10 If the **Plant** cannot be used because of a **Defect** to which this Clause applies, the **Defects Liability Period**, or if applicable the extended **Defects Liability Period**, shall be extended by a period equal to the period during which it cannot be used. Similarly the **Defects Liability Period**, or if applicable the extended **Defects Liability Period**, shall be extended by any period wherein the **Plant** cannot be used by reason of the **Contractor** attempting to put the **Plant** into such condition that it passes any relevant take-over procedures or any relevant performance test.

37.11 If for any reason the **Purchaser** does not allow the **Contractor** the necessary access to repair or correct any **Defect**, including permitting the **Contractor** to remove any defective **Materials**, the **Contractor** shall be relieved of his obligation to make such repair. However if the **Purchaser** merely defers the timing of such repair or corrective work for his own convenience, any additional cost which the **Contractor** incurs as a consequence of such deferment shall form part of the **Contract Price** irrespective of the underlying liability for the cost of the repair or corrective work. In no circumstance shall the **Purchaser** require the deferment under this Sub-clause of any repair or correction beyond the end of the relevant **Defects Liability Period** or any extension thereof.

37.12 Notwithstanding the provisions of Sub-clause 37.2, if any part of the **Plant** identified in Schedule 10 (Parts with limited working life) has a working life as stated therein of less than 365 days then the **Defects Liability Period** for such part shall be the working life so stated, and the provisions of this Clause shall be construed accordingly.

37.13 Nothing in this Clause shall be construed or take effect so as to impose on the **Contractor** any liability for any **Defect** in any work done or **Materials** supplied in accordance with Clause 9 (Assignment and subcontracting) or Clause 10 (Nominated Subcontractors) which shall exceed in terms of time, extent or otherwise that undertaken by the relevant **Subcontractor**.

38. Final Certificate

See Guide Note P (Limitations of Contractor's liability) and Appendix D (Sample Final Certificate)

38.1 Subject to the provisions of Sub-clauses 37.10, 38.2 and 38.3, as soon as the **Defects Liability Period** for the **Plant** has expired or the **Contractor** has made good all **Defects** that have within such period appeared in the **Plant** or a specified section thereof, in accordance with Clause 37 (Liability for Defects), whichever is the later, the **Project Manager** shall forthwith issue a certificate (a '**Final Certificate**') to the **Contractor** with a copy to the **Purchaser** stating that the **Plant** or specified section and any related work have finally been completed and the date of that completion.

38.2 If the provisions of Sub-clause 37.6 continue to apply to any part of the **Plant**, the **Project Manager** shall, as soon as the provisions of Sub-clause 38.1 are otherwise satisfied, issue a **Final Certificate** for the remainder of the **Plant** or specified section in which such part is included, provided that such part is then free from **Defects** which the **Contractor** is bound to make good under Clause 37. Such part shall thereafter be treated as if it were a separate specified section and shall be the subject of a separate **Final Certificate**.

38.3 If the repetition of any performance tests has been deferred in accordance with Sub-clause 35.10, no **Final Certificate** shall be issued in respect of the **Plant** or specified section to which such tests relate, until the **Plant** or specified section thereof has passed or is deemed to have passed such performance tests or until the **Contractor** has paid or agreed any liquidated damages that may become payable in respect of the failure of such tests.

38.4 The **Contractor** shall have no right or obligation to do any further work to any part of the **Plant** after a **Final Certificate** has been issued in respect of that part. The issue of the **Final Certificate** for the **Plant** and the **Works** as a whole or, where for any reason more than one **Final Certificate** is issued in accordance with this Clause, the issue of the last of the **Final Certificates**, which shall be identified as such, shall constitute conclusive evidence for all purposes and in any proceedings whatsoever between the **Purchaser** and the **Contractor** that the **Contractor** has completed the **Plant** and made good all **Defects** therein in all respects in accordance with his obligations under the **Contract**. No **Final Certificate** shall be conclusive as aforesaid if it or any other **Final Certificate** was issued in reliance upon any fraudulent act, misrepresentation or concealment.

38.5 Any dispute as to whether or not a **Final Certificate** should have been issued may be referred to an **Expert** in accordance with Clause 47 (Reference to an Expert).

39. Payment

See Guidance on compiling Schedules 18 (Cost elements, rates and charges) and 19 (Terms of payment)

39.1 Unless the **Contract** otherwise provides, payment of the **Contract Price** shall be made by monthly instalments.

39.2 If any instalment is to be paid upon the completion of a defined task or milestone, the **Contractor** shall only be entitled to apply for payment of such an instalment when he can provide evidence of completion of the task or achievement of the milestone as defined in Schedule 19.

39.3 The **Contractor** shall towards the end of each month submit to the **Project Manager** a statement showing:

(a) the **Contractor's** estimate of the expenditure that he will incur and the sums that will be due to him in the month succeeding the month in which the request is made, provided that in the first month the **Contractor** may submit his estimate for that month at its beginning;

plus

(b) the **Contractor's** actual expenditure and sums due to him for the month preceding the month in which the request is made;

less

(c) any amount previously certified by the **Project Manager** in respect of the **Contractor's** estimated expenditure and charges for the preceding month.

The **Contractor's** request for payment shall be supported by all relevant documentary evidence, including a statement showing the manner in which the total requested has been calculated.

39.4 Within fourteen days of the receipt of an interim statement, or in the case of the final statement within fifty-six days of its receipt, the **Project Manager** shall issue to the **Contractor** and the **Purchaser** a certificate for the instalment to which the statement relates. The certificate shall show the manner in which the sum certified has been calculated. The total certified shall comprise all sums listed in the **Contractor's** statement which, in the opinion of the **Project Manager**, are properly payable under the **Contract** and shall show separately any elements within the sums certified in respect of nominated **Subcontractors**. The **Project Manager** may in any certificate delete, correct or modify any sum previously certified by him as he shall consider proper.

Any such certificate issued by the **Project Manager** shall constitute notice from the **Purchaser** to the **Contractor** of the payment proposed to be made and the basis on which it has been calculated.

39.5 The sum certified in any certificate issued in accordance with Sub-clause 39.4 shall be due for payment on the date of such certificate. The **Purchaser** shall pay the amount so certified by a date (hereafter called the Final Date for Payment) which shall be twenty-eight days (or seventy days in the case of the final statement) after the date of receipt by the **Project Manager** of the **Contractor's** statement in accordance with Sub-clause 39.4.

39.6 If the **Purchaser** wrongfully fails to make payment, the amount unpaid shall bear interest compounded daily from the Final Date for Payment until it is received by the **Contractor**, at an annual rate which is two per cent above the **Agreed Rate** for the first month of delay. Such annual rate shall be increased by a further two per cent at the end of each further month of delay if the certified amount remains unpaid, up to the end of the third month. In the event that delay exceeds three months, any delay thereafter shall carry a rate of ten per cent above the **Agreed Rate**.

39.7 Without prejudice to any other remedy which he may have, the **Purchaser** shall be entitled to deduct from any payment due to the **Contractor** any sums which are due from the **Contractor** to the **Purchaser** under the **Contract**.

If the **Purchaser** intends to withhold any amount from or set off any amount against any payment which is due to the **Contractor** in accordance with the provisions of Schedule 19, including but not limited to sums that may be due from the **Contractor** to the **Purchaser** under the **Contract**, the **Purchaser** shall give notice to the **Contractor** not later than five days before the Final Date for Payment, specifying the amount he proposes to withhold and the ground for withholding payment. If there is more than one ground, such notice shall specify separately each ground and the amount attributable to it.

39.8 If proper certification of an instalment is not made in accordance with Sub-clause 39.4, or if the **Purchaser** does not make payment in full of the amount of an instalment certified in accordance with Sub-clause 39.4 reduced by the amounts stated in any notice issued in accordance with Sub-clause 39.7, the **Contractor** may give the **Purchaser** written notice of his intention to suspend performance of the **Works**. If such failure shall continue for seven days after the giving of such notice, then at any time thereafter and provided such failure is still continuing, the **Contractor** may suspend further performance of the **Works** until payment is made. All additional cost reasonably incurred by the **Contractor** as a result of such suspension and subsequent resumption shall form part of the **Contract Price** and the **Contract** shall be amended as may be agreed between the **Contractor** and the **Purchaser** or as may reasonably be necessary to accommodate the suspension of performance and the period of such suspension.

39.9 If the suspension of performance of the **Works** pursuant to Sub-clause 39.8 continues for a period of sixty days the **Contractor** may terminate his employment under the **Contract** by notice to the **Purchaser** and the provisions of Clause 42 (Termination by the Purchaser for convenience) shall apply as if the **Purchaser** had terminated the employment of the **Contractor**.

39.10 The **Contract Price** excludes Value Added Tax and to the extent that the tax is properly chargeable on the **Materials** and the **Works**, the **Purchaser** shall pay such tax as an addition to payments otherwise due to the **Contractor**.

39.11 If as a result of any audit pursuant to Clause 19 (Records and audits) or otherwise an error is discovered in the amount paid to the **Contractor**, then such error shall be corrected in the next payment due under the **Contract** or, following the making of the last payment, any finding of underpayment to the **Contractor** shall be remedied as quickly as possible and any finding of overpayment shall be remedied by the **Contractor** making the necessary re-payment within twenty-eight days of receipt of the relevant notification.

40. Contingencies and prime cost sums

See Guidance on compiling Schedule 18 (Cost elements, rates and charges)

40.1 If any sum is specified in Schedule 18 in respect of any contingency or additional work which may be required, the sum shall only be expended if the **Project Manager** issues an appropriate instruction by means of a **Variation Order**.

40.2 If a prime cost sum is specified in Schedule 18 for work to be done or **Materials** to be supplied by a nominated **Subcontractor** then, subject to the **Project Manager's** right to omit such work or **Materials** pursuant to Clause 16 (Variations), the **Project Manager** shall within a reasonable time give the **Contractor** instructions nominating the **Subcontractor** and, subject to Clause 10 (Nominated Subcontractors), the **Contractor** shall place a subcontract in accordance with such instructions.

41. Suspension of the Works

See Guide Note N (Suspension)

41.1 The **Project Manager** may suspend performance of all or any part of the **Works** by an order (herein called a '**Suspension Order**') given to the **Contractor** in writing and the **Contractor** shall thereupon do so until ordered to resume performance by the **Project Manager**. A **Suspension Order** shall be final, conclusive and binding on the **Contractor**. On receipt of a **Suspension Order**, the **Contractor** shall immediately advise the **Project Manager** of any aspects of the **Works** which need to be continued to maintain the safety and security of the **Plant** and **Works**.

41.2 If the **Contractor** is delayed in the performance of any of his obligations under the **Contract** by any **Suspension Order** given by the **Project Manager** (other than a **Suspension Order** given by reason of the **Contractor's** default), then any additional cost incurred by the **Contractor** as a result of such order shall form part of the **Contract Price**.

41.3 If by virtue of any **Suspension Order**, other than by reason of the **Contractor's** default, the **Contractor's** performance of any of his obligations is suspended for a continuous period of ninety days, then at any time thereafter and provided that such performance is still suspended, the **Contractor** may give notice to the **Project Manager** and to the **Purchaser** requiring that the **Project Manager** shall within fourteen days of such notice either withdraw the **Suspension Order** and instruct the **Contractor** to resume such performance, or issue a **Variation Order** excluding the suspended obligation from the **Contract**. If the **Project Manager** shall fail to do so then upon the expiry of such period of notice the employment of the **Contractor** shall unless otherwise agreed be deemed to have been terminated and thereupon the rights and obligations of the parties shall be as set forth in Clause 42 (Termination by the Purchaser for convenience).

41.4 If the **Project Manager** withdraws any **Suspension Order** then the **Approved Programme** and completion date shall be extended by the full period of suspension plus any reasonable additional time incurred by the **Contractor** to re-establish the activities of his own staff and labour, and those of his **Subcontractors**.

41.5 If any disagreement arises between parties under Sub-clause 41.2 or 41.4, the dispute may be referred to an **Expert** in accordance with Clause 47 (Reference to an Expert).

42. Termination by the Purchaser for convenience

See Guide Note O (Termination)

42.1 The **Purchaser** may at any time by written order to the **Contractor** (herein called a 'Termination Order') order the **Contractor** to cease the further carrying out of the **Works**.

42.2 The **Contractor** shall upon receipt of the Termination Order cease the execution of the **Works**, other than such work as the **Project Manager** may instruct. The **Contractor** shall upon completion of such work, remove any **Contractor's Equipment** from the **Site** and withdraw from the **Site** leaving it in a safe and tidy condition.

42.3 As soon as practicable after receipt of the Termination Order, the **Contractor** shall in particular:

(a) secure the cessation of work by his **Subcontractors** in accordance with and subject to the provisions of the preceding Sub-clause (so far as possible such cessation shall be secured by the exercise by the **Contractor** of such powers of termination, omission or cancellation as are available to him in the relevant subcontracts); and/or

(b) if so required by the **Purchaser**, assign to the **Purchaser** any subcontract designated by the **Purchaser** in preference to cancelling or terminating it.

42.4 Upon receipt of the Termination Order or on such later date as may be specified in the Termination Order, the **Contractor** shall assign to the **Purchaser** to the extent desired by the **Purchaser** all or any rights and titles held by the **Contractor** relating to **Materials** and the **Plant** together with the obligations connected therewith. The **Contractor** shall also deliver to the **Purchaser** all erection plans, schedules and **Documentation** and other data prepared by the **Contractor** or **Subcontractors** in connection with the **Works** and all **Documentation** supplied to the **Contractor** by or on behalf of the **Purchaser** in connection with the **Works**.

42.5 Within ninety days of the **Contractor's** withdrawal from the **Site**, or if the **Contractor** has not entered the **Site** before receipt of the Termination Order, then within ninety days of his receipt of the Termination Order, the **Project Manager** shall, subject to the provisions of Sub-clause 42.6, issue to the **Purchaser** and to the **Contractor** a certificate (hereinafter called a 'Termination Certificate') which shall state:

(a) the amount (if any) due to the **Contractor** under the **Contract** for the **Works** carried out prior to the receipt by the **Contractor** of the Termination Order;

(b) the amount due to any third party in respect of which the **Contractor** has (prior to the receipt by him of the Termination Order) properly and irrevocably entered into a commitment relating directly to the **Contract**; and

(c) the amount of any additional cost properly incurred by the **Contractor** for any work in connection with the termination authorised by the **Project Manager**.

The balance, if any, due to the **Contractor** shall be the sum of the amounts referred to in (a), (b) and (c) less that amount already paid to the **Contractor**. If the amount already paid to the **Contractor** exceeds the sum of (a), (b) and (c), the balance shall become due from the **Contractor** to the **Purchaser**.

If the **Contractor** cannot agree to any **Decision** of the **Project Manager** as to the amounts payable to the **Contractor**, the dispute may be referred to an **Expert** in accordance with Clause 47 (Reference to an Expert).

42.6 For the purpose of calculating the amounts of paragraphs (a), (b) and (c) of Sub-clause 42.5, account shall be taken in favour of the **Purchaser** of the realisable value of all things belonging to and purchased by the **Contractor** in accordance with the **Contract**.

42.7 As soon as practicable, the **Contractor** and the **Purchaser** shall provide the **Project Manager** with all such information and documents as he may reasonably require for the purpose of issuing the Termination Certificate.

42.8 If by the expiry of the period specified in Sub-clause 42.5 it is not possible for the **Project Manager** to issue a Termination Certificate by reason of any unresolved dispute between the **Contractor** and any **Subcontractor** or by reason of any other matter which prevents the ascertainment of the amounts referred to in paragraphs (a), (b) and (c) of Sub-clause 42.5, the **Project Manager** shall, at the expiration of such period, issue a provisional Termination Certificate which shall contain the best estimate that can then be made of any amounts referred to in the said paragraphs (a), (b) and (c) and of the resultant balance due. As soon thereafter as the ascertainment of the amounts referred to in the said paragraphs (a), (b) and (c) becomes practicable, the

Project Manager shall issue a Termination Certificate which shall operate as a correction or adjustment of the provisional Termination Certificate.

42.9 Payment of the balance due under any Termination Certificate shall be made between the **Purchaser** and the **Contractor** in accordance with any provisional or final Termination Certificate and the provisions of Clause 39 (Payment) shall apply as appropriate save that in such event the Final Date for Payment shall be fourteen days after the issue of the Termination Certificate.

42.10 Termination of the employment of the **Contractor** in accordance with this Clause shall be without prejudice to the continuing rights and obligations of the parties hereto with regard to the provisions of Clause 8 (Patent and other protected rights), Clause 20 (Confidentiality), Clause 45 (Disputes), Clause 46 (Adjudication), Clause 47 (Reference to an Expert) and Clause 48 (Arbitration) and, to the extent that the **Contractor** has carried out the **Works**, the provisions of Clause 30 (Care of the Works), Clause 31 (Insurance) and Clause 37 (Liability for Defects).

43. Termination for Contractor's default

See Guide Note O (Termination)

43.1 If the **Contractor** goes into liquidation (other than a voluntary liquidation for the purposes of reconstruction or amalgamation) or has an administration order made against him or carries on his business or any part of it under an administrator or receiver or manager for the benefit of his creditors or any of them, then without prejudice to any other rights or remedies, the **Purchaser** may forthwith by notice terminate the employment of the **Contractor** under the **Contract**.

43.2 If the **Contractor** is in default in that he:

(a) without reasonable cause wholly suspends or abandons the carrying out of the **Works** before completion thereof; or

(b) fails to proceed regularly and diligently with the **Works**; or

(c) commits any other material breach of the **Contract**;

then, without prejudice to any other rights or remedies which the **Purchaser** may possess, the **Project Manager** may notify the **Contractor** of such default and if the **Contractor** fails to commence and diligently pursue the rectification of the default within a period of fourteen days after receipt of notification, the **Purchaser** may by notice terminate the employment of the **Contractor** under the **Contract**.

43.3 Upon termination of the **Contractor's** employment under Sub-clauses 43.1 or 43.2:

(a) the **Purchaser** may himself or through others complete the **Works** and he or they may enter upon the **Site** and use the **Contractor's Equipment**, **Materials** and any other things whatsoever brought to the **Site** by the **Contractor** or which have become the property of the **Purchaser** pursuant to Sub-clause 25.1, and the **Purchaser** shall not be liable to the **Contractor** for any fair wear or tear or accidental damage that may occur to the **Contractor's Equipment**, **Materials** or other things;

(b) the **Contractor** shall, when so required by the **Purchaser**, forthwith deliver to the **Purchaser** all **Confidential Information** together with all **Documentation** and technical information prepared by the **Contractor** as referred to in Clause 21 (Documentation); and

(c) the **Contractor** shall, when so required by the **Purchaser**, assign to the **Purchaser** all his rights under any subcontracts.

43.4 If any item of **Contractor's Equipment** or of **Materials** which has not become the property of the **Purchaser** or any other thing is used for the completion of the **Works** in accordance with Sub-clause 43.3(a), then when that item or thing is no longer required for the purposes of the **Works**:

(a) the **Purchaser** shall notify the **Contractor** in writing of the availability of such item or thing and thereupon make such item or thing available for the **Contractor** to collect and remove forthwith and the **Contractor** shall so collect and remove such item or thing within twenty-one days of the notice;

(b) if the **Contractor** does not collect and remove the item or thing within a period of twenty-one days of its being made available in accordance with paragraph (a) hereof, the **Contractor** shall be deemed to have consented irrevocably to the disposal of the item or thing by the **Purchaser** and the **Purchaser** may thereafter dispose of that item or thing as he in his absolute discretion sees fit and any proceeds (less the cost of such disposal) shall be paid to the **Contractor**. The **Contractor** shall be liable to pay to the **Purchaser** those costs of disposal not recovered by the **Purchaser** from any proceeds. The **Project Manager** shall certify the amounts payable to or by the **Contractor** pursuant to this Sub-clause and the provisions of Clause 39 (Payment) shall apply thereto as appropriate save that the Final Date for Payment shall be fourteen days after the issue of the **Project Manager's** Certificate.

43.5 Upon the termination of the **Contractor's** employment under this Clause, the **Purchaser** shall have no obligation to make any further payment beyond that already certified by the **Project Manager** other than as may be certified in accordance with Sub-clauses 43.6 and 43.8.

43.6 Within ninety days of the date of the termination of the **Contractor's** employment or the completion of the **Works** in accordance with Sub-clause 43.3 (a) (including for the avoidance of doubt all testing and remedying of defects, such that the total cost to be incurred by the **Purchaser** has been incurred) the **Project Manager** shall, subject to the provisions of Sub-clause 43.8, issue to the **Purchaser** and the **Contractor** a certificate (hereinafter called a 'Default Certificate') which shall set out a full statement of account including:

(a) all sums due to the **Purchaser** from the **Contractor** including without limitation any cost incurred by the **Purchaser** in completing the **Works** in accordance with Sub-clause 43.3(a) and which is in addition to that which the **Purchaser** would have incurred if the **Contractor** had completed the **Works** in accordance with the **Contract**; and

(b) all sums due to the **Contractor** in respect of work undertaken by the **Contractor** prior to the termination of his employment other than any such work of a temporary nature necessitated by such termination.

Having allowed for all previous payments made to the **Contractor** the Default Certificate shall state the balance due to or from the **Contractor**.

43.7 As soon as practicable, the **Contractor** and the **Purchaser** shall provide the **Project Manager** with all such information and documents as he may reasonably require for the purpose of issuing the Default Certificate.

43.8 If by the expiry of the period specified in Sub-clause 43.6, it is not possible for the **Project Manager** to issue a Default Certificate by reason of any matter which prevents the ascertainment of the amounts referred to in Sub-clause 43.6, the **Project Manager** shall, at the expiration of such period, issue a provisional Default Certificate which shall contain the best estimate that can be made of any amounts referred to in the said Sub-clause and of the resultant balance due. As soon thereafter as the ascertainment of the amounts referred to in the said Sub-clause becomes practicable, the **Project Manager** shall issue a final Default Certificate which shall operate as a correction or adjustment of the provisional Default Certificate.

43.9 If the **Contractor** cannot agree to any **Decision** of the **Project Manager** as to the amounts payable by or to the **Contractor**, the dispute may be referred to an **Expert** in accordance with Clause 47 (Reference to an Expert).

43.10 Payment of the balance due under any Default Certificate shall be made between the **Purchaser** and the **Contractor** in accordance with any provisional or final Default Certificate and the provisions of Clause 39 (Payment) shall apply as appropriate save that the Final Date for Payment shall be fourteen days after the issue of the Default Certificate.

43.11 Termination of the employment of the **Contractor** in accordance with this Clause shall be without prejudice to the continuing rights and obligations of the parties hereto with regard to the provisions of Clause 8 (Patent and other protected rights), Clause 20 (Confidentiality), Clause 45 (Disputes), Clause 46 (Adjudication), Clause 47 (Reference to an Expert) and Clause 48 (Arbitration) and, to the extent that the **Contractor** has carried out the **Works**, the provisions of Clause 30 (Care of the Works), Clause 31 (Insurance) and Clause 37 (Liability for Defects).

44. Limitation of liability

See Guide Note M (Liability for Defects) and Guide Note P (Limitation of Contractor's liability). For projects outside the United Kingdom, see Guide Note U (Projects outside the United Kingdom)

44.1 Neither the **Contractor** nor the **Purchaser** shall be liable to the other for:

(a) wastage, loss or contamination during its use in the **Plant** of any process consumable which shall be deemed to include feedstocks, chemicals, biochemicals, catalysts and utilities; and

(b) loss or deferment of anticipated or actual profit, loss of revenue, loss of use, loss of production, business interruption or any similar damage or for any consequential or indirect losses of any other kind resulting from or arising out of or in connection with the **Works** or the performance thereof or any act or omission relating thereto howsoever caused;

except for:

(i) recoveries in respect thereof obtained as a result of the insurance, etc. under Clause 31 (Insurance);

(ii) the **Contractor's** obligations with respect to the payment of liquidated damages for delay under Sub-clause 15.1;

(iii) the **Contractor's** obligations with respect to liquidated damages for failure to pass performance tests under Sub-clause 35.9; or

(iv) any express provision to the contrary in the **Contract**.

44.2 Except in the case of termination of the **Contractor's** employment under the terms of Clause 43 (Termination for Contractor's default), the liability of either party to the other for any breach of contract shall be limited to the expenses, charges, damages and reimbursements expressly provided in the **Contract**. Nothing in the **Contract** shall in any way be interpreted as affecting or limiting any liability which the **Contractor** may have under the Consumer Protection Act 1987 or in respect of personal injury or death caused by the negligence of the **Contractor** (as defined in Section 1 of the Unfair Contract Terms Act 1977).

44.3 The total aggregate liability of the **Contractor** to the **Purchaser**, other than liability arising under Sub-clause 8.2 and Clauses 30 (Care of the Works) and 31 (Insurance), shall not exceed the amount stated in the Form of Agreement.

Any sums recovered by the **Contractor** from any **Subcontractor** under or pursuant to any subcontract, whether as damages, default payments, or costs for breach of contract or under settlements in respect of alleged breach of contract (such settlements to be made only with the consent of the **Purchaser**, which consent shall not be unreasonably refused) shall be paid in full to the **Purchaser**.

44.4 Any exclusion or limitation of liability under the **Contract** shall as between the parties exclude or limit such liability in contract and in tort including negligence or otherwise.

45. Disputes

See Guide Note Q (Dispute resolution)

45.1 The **Purchaser** and the **Contractor** shall use all reasonable endeavours to avoid disputes both between themselves and with third parties including, but not limited to, **Subcontractors**.

45.2 In order to avoid the development of disputes and to facilitate their clear definition and early resolution, the procedures set out in Clauses 45 (Disputes), 46 (Adjudication), 47 (Reference to an Expert) and 48 (Arbitration) shall be applied as appropriate. The **Purchaser** and the **Contractor** undertake to use reasonable endeavours to avoid the escalation of problems into 'disputes' as defined in Sub-clause 45.4. However the parties acknowledge and agree that this undertaking shall not prejudice either party's rights under Clause 46 to refer any dispute or difference to adjudication at any time if that party so wishes notwithstanding the definition of 'dispute' in Sub-clause 45.4.

45.3 If the **Contractor** is dissatisfied with any **Decision** or valuation of the **Project Manager**, or of any person to whom the **Project Manager** may have delegated any of his authority or responsibility, or if the **Purchaser** or the **Contractor** is dissatisfied with any other matter arising under or in connection with the **Contract**, either party may at any time refer such dissatisfaction to the **Project Manager** giving full details of the nature of the matter. The **Project Manager** shall give a written **Decision** on the matter (giving the reasons for such decision) to the **Purchaser**, the **Contractor** and the **Contract Manager** within twenty-eight days of such reference to him.

45.4 The **Purchaser** and the **Contractor** agree that no matter shall constitute, nor be said to give rise to, a dispute, which shall include any difference, unless the same has been referred to the **Project Manager** under Sub-clause 45.3 and:

(a) the **Project Manager** has failed to give his **Decision** on the said matter within the prescribed time; or

(b) a **Decision** given within the prescribed time is either unacceptable to the **Purchaser** and/or the **Contractor** or has not been implemented within twenty-one days of the said **Decision**;

and, as a consequence, either the **Purchaser** or the **Contractor** has served notice setting out the nature of the dispute (hereinafter called a 'Notice of Dispute') on the other and on the **Project Manager**. For the purposes of the performance of the **Works** and all matters arising out of or in connection with the **Contract**, the word 'dispute' shall be construed in accordance with this Sub-clause 45.4.

45.5 Notwithstanding the existence of any dispute or any reference to the **Project Manager** under Sub-clause 45.3, the **Purchaser** and the **Contractor** shall continue to perform their obligations under the **Contract**.

45.6 Subject to any other provisions of the **Contract** the parties shall attempt to negotiate a settlement of any dispute in good faith.

45.7 If a dispute cannot be resolved by negotiation the parties may by agreement refer it to mediation in accordance with the procedures of the Centre for Dispute Resolution (CEDR), the Chartered Institute of Arbitrators (CIArb) or some other body.

45.8 No **Decision**, opinion, direction or valuation given by the **Project Manager** shall disqualify him from being called as a witness and giving evidence before a third party, an **Expert**, adjudicator or arbitrator on any matter whatsoever relating to a dispute.

46. Adjudication

See Guide Note Q (Dispute resolution)

46.1 This Clause shall only apply to disputes arising under a construction contract as defined in the Housing Grants, Construction and Regeneration Act 1996, or any amendment or re-enactment thereof.

46.2 Notwithstanding any provision in these General Conditions for a dispute to be referred to an **Expert** in accordance with Clause 47 (Reference to an Expert) or to Arbitration in accordance with Clause 48 (Arbitration), either party shall have the right to refer any dispute or difference (including any matter not referred to the **Project Manager** in accordance with Sub-clause 45.3) as to a matter under or in connection with the **Contract** to adjudication and either party may, at any time, give notice in writing to the other of his intention to do so (hereinafter called a 'Notice of Adjudication'). The ensuing adjudication shall be conducted in accordance with the edition of the 'Adjudication Rules' (the 'Rules') published by IChemE current at the time of service of the Notice of Adjudication.

46.3 Unless the adjudicator has already been appointed, he is to be appointed to a timetable with the object of securing his appointment within seven days of the service of the Notice of Adjudication. The appointment of the adjudicator shall be effected in accordance with the Rules.

46.4 The adjudicator shall reach his decision within twenty-eight days of referral or such other longer period as may be agreed between the parties after the dispute has been referred.

46.5 The adjudicator may extend the period of twenty-eight days by up to fourteen days with the consent of the party by whom the dispute was referred.

46.6 The adjudicator shall act impartially.

46.7 The adjudicator may take the initiative in ascertaining the facts and the law.

46.8 The decision of the adjudicator shall be binding until the dispute is finally determined by legal proceedings, by arbitration or by agreement.

46.9 The adjudicator shall not be liable for anything done or omitted in the discharge or purported discharge of his functions as adjudicator unless the act or omission is in bad faith. Furthermore, any employee or agent of the adjudicator acting in connection with the carrying out of the adjudication shall be similarly protected from liability.

47. Reference to an Expert

See Guide Note Q (Dispute resolution)

47.1 Subject to the provisions of Clause 45 (Disputes), any dispute which in accordance with the **Contract** is a matter which may be referred to an **Expert** in accordance with this Clause shall be so referred upon one party giving notice thereof to the other. Furthermore, if the parties so agree any other dispute arising out of or in connection with the **Contract** may be referred to an **Expert**.

The **Expert** shall be agreed between the parties or, in the absence of such agreement within twenty-one days of the service of the notice under this Clause, shall be appointed by the President for the time being (or a Past President) of IChemE upon the application of either party.

If the **Expert** declines the appointment, or after the appointment is removed by order of a competent court, or dies or otherwise becomes incapable of acting and the parties do not within one month of the vacancy arising fill the vacancy by agreement, then either party may apply to the President for the time being (or a Past President) of IChemE to appoint another **Expert** to fill the vacancy.

Any such reference shall be conducted in accordance with the edition of the 'Rules for Expert Determination' published by IChemE current at the time of the **Expert's** appointment unless the parties otherwise agree prior to the appointment of the **Expert**.

47.2 The **Expert** shall decide all disputes referred to him as an expert and not as an arbitrator. Any decision of the **Expert** may revise or overrule any **Decision** of the **Project Manager** other than any **Decision** expressly stated in the **Contract** to be final, conclusive and binding.

47.3 The powers of the **Expert** to decide disputes shall include decisions as to factual issues and the interpretation of the **Contract**.

47.4 The **Purchaser** and the **Contractor** shall afford the **Expert** every assistance in deciding any dispute referred to him and shall give him access to the **Site** and to their premises and shall provide him with any information and **Documentation** that he may reasonably require.

47.5 The **Expert** shall have power by his decision to fix the reasonable amount of his fees in connection therewith and they shall be borne in equal shares between the **Purchaser** and the **Contractor**.

47.6 If the **Contractor** has complied with any **Decision** of the **Project Manager** which is overruled by the **Expert**, and has thereby suffered or incurred any loss or additional expense in the performance of the **Contract**, then provided that the **Contractor** has notified the **Project Manager** that he disputes such **Decision** before complying with it, the **Expert** may in his discretion award the **Contractor** a reasonable sum in respect of such loss or additional expense, notwithstanding that the overruling of the **Decision** would not otherwise entitle the **Contractor** to any additional payment under the **Contract**.

47.7 The **Purchaser** and the **Contractor** hereby agree to be bound by any decision of the **Expert** under this Clause, which decision shall be final, conclusive and binding, and shall comply with any direction given therein and shall not question the correctness of any such decision or direction in any proceedings.

47.8 Neither the **Purchaser** nor the **Contractor** shall be entitled to suspend performance of the **Contract** by reason of the reference of a dispute to an **Expert**.

47.9 Any dispute which is referred to an **Expert** ceases to be referable to arbitration under Clause 48 (Arbitration).

48. Arbitration

See Guide Note Q (Dispute resolution)

48.1 Any dispute, other than failure to give effect to a decision of an adjudicator, which has not been settled whether or not following the procedures in Clause 45 (Disputes) or Clause 46 (Adjudication) or which has not been referred to an **Expert** in accordance with Clause 47 (Reference to an Expert) shall be referred to a single arbitrator.

48.2 The arbitrator shall be appointed by agreement between the parties.

If the parties fail to appoint an arbitrator within one month of either party serving on the other party notice in writing to concur in the appointment of an arbitrator, then the dispute shall be referred to a person to be appointed on the application of either party, by the President for the time being (or a Past President) of IChemE.

48.3 The arbitration shall be conducted in accordance with the procedure set out in the edition of the 'Arbitration Rules' published by IChemE current at the time of appointment of the arbitrator.

The arbitrator shall have the full power to open up, review, revise or overrule any **Decision** of the **Project Manager**, other than any **Decision** expressly stated in the **Contract** to be final, conclusive and binding.

Neither party shall be limited in arbitration to the evidence or arguments put to the **Project Manager** or to an adjudicator pursuant to Sub-clauses 45.3 and 46.2 respectively.

48.4 The **Purchaser** and the **Contractor** hereby agree to be bound by any decision, award or ruling of an arbitrator under this Clause and shall comply therewith. The **Purchaser** and the **Contractor** hereby agree to exclude any right to appeal to the Court under the Arbitration Act 1996 or otherwise to seek to challenge the arbitrator's decision, award or ruling in any court proceedings whatsoever.

48.5 If the **Contractor** has complied with any **Decision** of the **Project Manager** which is overruled by the arbitrator and has thereby suffered or incurred loss or additional expense in the performance of the **Contract**, then provided that the **Contractor** has notified the **Project Manager** that he disputes such **Decision** before complying with it, the arbitrator may in his discretion award the **Contractor** a reasonable sum in respect of such loss or additional expense, notwithstanding that the overruling of the **Decision** would not otherwise entitle the **Contractor** to any additional payment under the **Contract**.

48.6 Neither the **Purchaser** nor the **Contractor** shall be entitled to suspend performance of the **Contract** by reason of the reference of a dispute to arbitration.

Guidance on compiling the Agreement, its Annex, the Specification and Schedules

The purpose of this section is to provide an *aide-mémoire* to help the user of this handbook to prepare a complete set of documentation to accompany the Conditions of Contract. This guidance has been drafted in the light of English Law current at the time of publication.

It is not intended to give a comprehensive or definitive statement of the commercial or legal position, nor does it form part of the Conditions or of any contract.

Guidance on completing the Agreement

The amount to be entered in Paragraph 3 will depend on the value of the property involved and the Purchaser's preference for the cost of insurance to be paid by the Contractor, who will have allowed for it in the Contract Price.

The amount of the deductible for the Purchaser's CAR policy to be entered into Paragraph 4 will be determined having regard to the fact that if set too high the Contractor will be likely to carry insurance to reduce the deductible he bears himself, passing the cost of the extra insurance on to the Purchaser through his management fee.

When completing paragraph 5 of the Agreement, the total aggregate liability of the Contractor to the Purchaser, it should be borne in mind that the amount entered is a matter for negotiation between the parties. Matters to be considered include the Contractor's liability under Sub-clause 37.3 and for liquidated damages.

Paragraph 7 requires the date of the Contractor's tender to be inserted. This sets an appropriate base-date for the matters referring to this date in the sub-clause.

The date for the commencement of the Works is needed in Paragraph 8 because it could be earlier or later than the date of the Agreement, and is relevant to the periods for completion in Schedule 11 (Times of completion).

The naming of the Project Manager in Paragraph 9 who is not necessarily an employee of the Purchaser is required because of the Contractor's obligations to communicate with the Purchaser through the Project Manager (see Sub-clause 11.8, etc.) Note that the definition of Project Manager in Clause 1 (Definition of terms) states that the Project Manager is an individual, not a committee.

The naming of the Contract Manager in Paragraph 10 is required for the same reason. He may be regarded by the Purchaser to be of special importance because of his particular experience. It is for this reason that Sub-clause 12.1 provides that the Contract Manager will not be replaced without the consent of the Project Manager.

Guidance on the Annex to the Agreement

The final sentence of Paragraph 1 of the Agreement, following the list of documents which form the Contract, states:

'For the purpose of identification, the contents of the Contract, including the number of pages in each part, are listed in the Annex to this Agreement attached hereto.'

Clarity as to what documents form part of the Contract is essential, as is knowledge as to the extent of those documents—for example, has the last page been lost?

The Annex to the Agreement is therefore intended to be a simple list of the various documents which form part of the Contract, with the total number of pages in each document stated. A unique reference number on each page, the Purchaser's contract number for instance, or name of the project or contract will also be helpful in giving clarity as to the identity of those documents.

The Annex itself should be clearly identified as such, and be attached to the Agreement.

Guidance on compiling the Specification

See Sub-clauses 3.1, 3.4, 13.1, 32.1, 33.1, 35.2, 35.4, 36.2, 37.1 and 38.1

Every process plant contract needs a Specification, the technical definition of the Plant that the Contractor has agreed to design and construct. The more precise and detailed the Specification, the greater the Contractor's understanding of his commitment.

In a price-based contract a comprehensive detailed Specification is necessary. It allows the Contractor's bid to be fully and firmly based and provides the measure against which the adequacy of the finished Plant and its performance, the Contractor's execution of his obligations and any variations to the Plant may be checked easily.

A reimbursable contract gives the parties much more freedom. The Specification may vary from a very brief statement outlining the type and capacity of the Plant to a detailed description of the equipment required. In general terms the less precise the initial Specification the greater the degree to which the Contractor will need to be paid on a reimbursable basis. However if any part of the Contractor's supply (material, labour or services) is to be covered by a lump sum then the initial Specification must be appropriately detailed.

If the initial Specification is not fully comprehensive it should be progressively completed as the project proceeds. It is then important to identify a stage in the project when the necessary information will be available and at that time to complete the Specification. In its final form the Specification must include a full description of the Purchaser's requirements for the Plant together with a list of all the design standards and codes of practice etc. to which the Plant is to be built, and should cover such matters as corrosion allowances and the design life of the different elements.

A Purchaser who is subject to European Directives on Public Procurement and the national law implementing them is reminded of the provisions therein relating to specifications and standards.

In cases where codes of practice, guidelines or standards are less well defined or where guides to good manufacturing practice are referenced, the Contractor and the Purchaser must take great care to provide sufficient detail in the Specification to avoid any ambiguity in the definition of the standards to which the Plant is to be constructed. In particular, where a third party inspectorate will eventually rule on the sufficiency and/or suitability for purpose of the Plant, then this must also be clearly stated in the Specification. Where the acceptance of the Plant by a third party inspector may have to be based on regulations which are open to interpretation, the Purchaser has a right to expect the Contractor to design a Plant which can comply with any reasonable interpretation.

The final Specification should be a document which is completely consistent and which:

(a) incorporates all technical details as finally agreed;
(b) identifies the engineering standards that are to be used;
(c) identifies sections, if any, of the Plant which the Purchaser requires to be completed, taken over or tested for performance as separate sections.

If early availability of one or more sections independently of the others is of special importance to the Purchaser, such sections should be clearly identified.

The capability of the Plant to make the desired product or products from specified feedstocks will form a principal feature of the Specification, while the required hourly or daily capacity should be demonstrated in a performance test. It should be recognised that some processes can involve an 'ageing' factor—catalyst activity for example—and the designer should choose an economic working life for any components or materials that have to be replaced periodically. Similarly there may be components of equipment or machinery which have been designed to work for only a limited life. These should be listed in Schedule 10 (Parts with limited working life) and reference made to the Schedule in the Specification. It then becomes important to specify the performance that should be achieved when the components or materials are new and fully active and how the consequences of changes over time are to be measured and reflected in the evaluation of performance.

The Purchaser may also wish to specify the process performance at the end of the working life of the material or catalyst that requires the longest interruption to production to replace, but it is unlikely that this could be demonstrated by a performance test, because of the time factor. In any case the conditions under which the performance tests are to be carried out, if in any way exceptional, should be stated (see Sub-clause 35.4).

Finally, particularly where the work involves the modification or extension of an existing plant, the Specification may need to take special account of the work to be done to that plant. This obviously brings a whole range of additional considerations into play, such as the possible need to maintain production, the mechanical condition of the original plant (which is in effect being supplied by the Purchaser), the interface between the original plant and the new equipment, etc. Consideration may also need to be given to the question of whether any of the general conditions, such as Clause 30 (Care of the Works) for example, need to be qualified by a Special Condition.

Guidance on compiling Schedule 1: Description of the Works

See Clauses 3 (Contractor's responsibilities), 7 (Statutory and other obligations), 16 (Variations), 23 (The Site), 26 (Health, Safety and Environment), 38 (Final Certificate), 42 (Termination by the Purchaser for convenience) and 43 (Termination for Contractor's default), Guidance on compiling Schedule 3 (Responsibilities of Purchaser), Guide Note C (Intellectual property, know-how, confidentiality and information) and Guide Note G (Site services and working conditions)

Whilst the Specification defines the Plant, Schedule 1 should specify the Site and the services to be provided by the Contractor including project evaluation if required, design work, the procurement and construction of the Plant and any temporary works, and any other work to be carried out by the Contractor. The Schedule should include all the information that the Contractor requires to carry out the Works.

The guiding principle in drawing up this Schedule and Schedule 3 (Responsibilities of Purchaser) is that they should between them make clear the dividing lines between the respective responsibilities and scope of work of the Contractor and Purchaser, thereby defining the limits of the Contractor's responsibilities under the Contract.

The site description should include:

(a) a written description which covers location, map references, what the Site comprises;
(b) plans of the Site including maps and layout, clearly indicating any hazardous, environmentally sensitive or 'no go' areas.

Where the Site is an existing plant, it is clearly important for Schedule 1 to provide comparatively detailed information concerning the topography and characteristics of the Site, including information on former uses, buildings and facilities located thereon.

All documentation necessary to complement the information contained in the Schedule should be listed and attached to the Schedule.

A very large number of activities concerning the provision of information and designs, the supply of materials and of construction services enter into the execution of any process plant project. These are outlined in the check-list at the end of this guide note. This list is only given as an *aide-mémoire*; it is not complete, nor are the groupings of headings and other details always applicable.

Note that every applicable item must fall within the responsibility of either the Purchaser or the Contractor. There are only two parties to the Contract, the Purchaser and the Contractor. Whatever arrangements either of these parties may have with others, resulting in the provision of certain services by them, those arrangements are subject to separate agreements and will form no part of the Contract. Schedules 1 and 3 should never attempt to give a full list of the contractual responsibilities of both parties, but should only state in detail the responsibilities of one party.

Schedule 1 describes the project and the extent and nature of the Contractor's contribution to the project in terms of process technology, engineering and other office services (including procurement services), equipment items and other Materials, and field supervision, labour, tools and plant, temporary works, etc. Frequently, and particularly when the project is at an early stage, the description of the project and/or the Contractor's supply may have to be qualitative rather than quantitative. For example, it may not even be possible to specify the process or plant design capacity, or even the exact location, as the determination of this may be part of the Contractor's work. In this case it is important to identify the stage of the Contractor's work at which a Specification of the Plant is to be agreed. It is good practice at this stage for the Contractor to provide a final estimate based on this agreed specification and use this as the cost control estimate.

However, it is still important that the precise extent of the Contractor's supply is adequately defined in Schedule 1 and that the methods by which the Contractor's charges for services rendered are to be calculated are clearly spelled out in Schedule 18 (Cost elements, rates and charges). (See the Introductory Notes for the case where part of the Contractor's supply is to be covered by a lump sum.)

One reason for the adoption of a reimbursable format may be that the Purchaser has a general view of what he requires, but needs the Contractor to carry out studies which could include alternatives for the Purchaser to review before choosing one on which to proceed. Any such requirement should be listed in Schedule 1.

As regards process design, this may be provided by a licensor under a direct contractual arrangement with the Purchaser. For the purpose of the Contract in this case the process information is the responsibility of the Purchaser. If detailed process engineering is meant to be provided by the Contractor, then it is necessary to define exactly what this comprises in terms of equipment data, detailed flow diagrams, instrumentation details and so on. Schedule 3 should set out the full and precise extent of the process information to be provided by the Purchaser; the Contractor would then do whatever else is necessary, including detailed design, based on this information. Sometimes, however, only the basic process design is supplied by the Purchaser, perhaps from his own research data; under these circumstances the Contractor will have to do whatever else is necessary, but he will start at an earlier point in the sequence of events leading to the final design. The exact starting points for the Contractor should be stated unambiguously, as should the Contractor's responsibility, total or partial, for process guarantees. Great care is needed in writing these critical definitions. If the Contractor offers a Plant complete with process design then he takes full responsibility for this aspect even though the relevant information comes to him from a separate licensor (see Guide Note C (Intellectual property, know-how, confidentiality and information)).

When technology is licensed from a third party, whether by the Purchaser directly or by the Contractor as part of his contract, particular care must be taken to define in the relevant schedule the role of licensor's personnel in checking and supervising design, construction, advising on and/or supervising start-up and in advising on or supervising any performance tests.

Whilst it is normal for Purchasers to take the lead in commissioning, sometimes the Contractor is required to provide additional services after the Plant start-up and during the early months of its operation; for instance, he may have to carry out an extensive operator training programme. In such cases the arrangements and responsibilities involved should be defined and detailed in Schedule 9 (Training by Contractor) which should include procedures, key staff to be trained, and duration and timing of the training. This training requirement should extend beyond plant operation to include such matters as health and safety, quality assurance and maintenance procedures.

'Qualification procedures' and 'validation procedures' are becoming increasingly important, particularly when the Purchaser requires the Contractor to achieve a specific standard of quality either defined in a national or international standard or when a third party regulating body insists on prescribed and highly-developed testing procedures to ensure that the Plant operates satisfactorily within the tolerances stated in the Specification. Qualification and validation apply to all phases of the development of a plant, including design, installation, operation and process. It is normal for the Contractor to be required to carry out the qualification up to and including operation of the plant but usually excluding qualification of the process, particularly if the process or technology is owned by the Purchaser. Schedule 6 (Quality assurance and validation) should outline the Purchaser's requirements.

The details about site access necessary in Schedules 1 and 3 will vary depending on whether the Site is undeveloped or part of an existing facility. In Sub-clause 23.2 it is assumed that the Site belongs to the Purchaser and that if necessary the Purchaser will arrange that the Site has a suitable right of access from the public highway.

In this context access could mean a clear access route over land without any road paving or railway track, in which case Schedule 1 could include the construction of an access way as part of site preparation. Alternatively the Purchaser could have arranged for a third party to construct the access way ready for use by the Contractor. In any case it will be essential for the Contractor to ensure that the access way has the capacity required, both in terms of weight and clearances for large loads on transporters, for the passage of the largest loads to be brought on to the Site. The Contractor will have to take account of any limitations (for example, maximum size of load that can be transported around the Site) in estimating—for example, by allowing for a greater amount of site assembly than he would ideally like—for while it is the Purchaser's duty to provide a suitable right of access, just what is suitable will be for definition having regard to the equipment, materials and labour to be brought onto the Site.

Similar issues apply to the route to the Site and to the use of utilities on the Site. Sub-clause 23.2 (second paragraph) makes these wholly the Contractor's risks. Sub-clause 23.3 makes the Contractor liable for any damage along any route to the Site.

If the Site is either part of or immediately adjacent to an existing installation, there may be specific constraints upon access arising from the Construction (Design and Management) Regulations 1994 (the 'CDM Regulations'). For example, the Purchaser, or another contractor engaged by the Purchaser, may be the designated 'Principal Contractor' with responsibility for the control of access. Details of any such constraints must be set out in Schedule 1.

Check-list

(1) Project evaluation:
- (a) market surveys;
- (b) raw material availability;
- (c) waste disposal facilities (see 3(e));
- (d) site conditions—climate, access, ground condition, labour situation;
- (e) plant capacity and optimisation.

(2) Preliminary activities, local surveys and site investigations.

(3) Definition of project data:
- (a) process requirements;
- (b) raw materials, feedstocks, utilities;

(c) products, intermediate products and by-products;
(d) effluents, with limitations on discharges and emissions (see Schedule 5 (Environmental protection and waste disposal));
(e) battery limits (see the Specification);
(f) battery limit conditions for (c), (d) and (e) above;
(g) statutory requirements.

(4) Process design:
(a) basic flowsheets;
(b) process calculations;
(c) heat and mass balances;
(d) hazardous materials data sheets;
(e) process design manual (see Schedule 2 (Documentation)).

(5) Detailed chemical engineering design:
(a) equipment data;
(b) detailed flow diagrams;
(c) instrumentation and control data;
(d) process control computer interfacing;
(e) materials of construction.

(6) Detailed mechanical and structural engineering design (see also (8) below):
(a) equipment list;
(b) equipment sketches or drawings;
(c) equipment specifications;
(d) layout;
(e) piping arrangements and details;
(f) models;
(g) line diagrams (piping, mechanical, etc.);
(h) structural steelwork;
(i) insulation and painting requirements;
(j) engineering standards;
(k) labelling and numbering system;
(l) heating, ventilation and air conditioning system design;
(m) room data sheets including environmental standards.

(7) Electrical installation design:
(a) single line diagrams;
(b) hazardous area electrical classification;
(c) list of electric motor drives;
(d) power distribution scheme;
(e) incoming power supplies;
(f) electrical sub-station and transformer specification;
(g) sub-distribution board specification;
(h) uninterruptible power supply specification;
(i) emergency power supply scheme;
(j) emergency generator specification;
(k) wiring and cable standards;
(l) motor control centre design.

(8) Control and instrumentation:
(a) control and instrumentation philosophy;
(b) single line diagrams;
(c) control loop diagrams;
(d) computer hardware and software standards;
(e) control logic diagrams;
(f) instrument lists;
(g) data collection and recording requirements;
(h) fire detection system;
(i) security and personnel access system;
(j) telephones and communications;
(k) gas detection system;
(l) operator/control interface.

(9) Civil engineering and building design:
(a) outline foundation and building drawings;
(b) detailed design of foundations, structures, buildings;
(c) roads, trenches, earthworks;
(d) specifications, bills of quantities;
(e) documentation for planning approval;
(f) documentation and calculations for building regulations approvals;
(g) drainage (storm, foul and process);
(h) effluent holding and treatment system.

(10) Project co-ordination procedures:
(a) project programme;
(b) reporting;
(c) cost report;
(d) procurement;
(e) communication, documentation, distribution and transmittal procedures (see Sub-clause 11.7 and Guide Note A (Communications));
(f) document identification and numbering system;
(g) plant design and performance qualification;
(h) quality assurance standards to be adopted (see Schedule 6 (Quality assurance and validation));
(i) validation.

(11) Supply of plant, equipment and other materials, packaged units, chemicals and catalysts (see Schedule 7 (Subcontracting)):
(a) enquiring and purchasing;
(b) expediting;
(c) inspections—statutory and non-statutory;
(d) factory and site acceptance testing;
(e) shipping services, transport to Site;
(f) test certification;
(g) approved vendors lists;
(h) special tools and spare parts;
(i) supply of chemicals and catalysts.

(12) Construction aspects (see Guide Note G (Site services and working conditions)):
(a) access to Site;
(b) off-loading, reception and storage of materials;
(c) construction manpower histograms identifying trade disciplines;
(d) Contractor's equipment and tools;
(e) supervision at various levels (see Schedule 8 (Contractor's named personnel));
(f) services and utilities;
(g) subcontracts;
(h) civil engineering works;
(i) Site administration including normal working hours (see Sub-clause 28.5);
(j) Site conditions;
(k) Site personnel policy;
(l) Site safety policy (note that the Purchaser's responsibility for providing copies of all his rules affecting access, safety, working methods or industrial relations should be stated in detail in Schedule 3 (Responsibilities of Purchaser));
(m) precautions to be taken regarding contaminated soil, adjacent premises, etc.

(13) Preparation for start-up: (see Guide Note S (Completion, taking over, testing and start-up)):
   (a) operating manuals;
   (b) maintenance manuals;
   (c) mechanical testing (including pressure testing, testing of lifting beams and other equipment, etc.) (see Schedule 14 (Criteria for completion of construction) and Schedule 15 (Take-over procedures));
   (d) plant operators;
   (e) supervision;
   (f) specialists;
   (g) process materials, lubricants;
   (h) training for all personnel (see Schedule 9 (Training by Contractor));
   (i) spares availability and store keeping;
   (j) performance qualification documentation;
   (k) safety system tests.

(14) Start-up, Plant operations:
   (a) operating labour;
   (b) supervision;
   (c) specialists;
   (d) analytical services;
   (e) routine maintenance;
   (f) alterations, construction services;
   (g) performance tests;
   (h) Plant performance studies;
   (i) emergency procedures;
   (j) Plant shutdown procedures.

(15) Safety (see Schedule 4 (Health and Safety)):
   (a) safety review records;
   (b) risk assessment/safety case;
   (c) assessment of hazardous substances (COSHH) or local equivalent;
   (d) chemical data sheets;
   (e) Hazard Analysis Assessment (HAZAN) and Hazard Operability Study (HAZOP);
   (f) safety review procedure (including major hazard site assessment);
   (g) disaster plan;
   (h) safety plan and risk assessment (Construction (Design & Management) Regulations or local equivalent).

(16) Other activities:
   (a) export and import licences;
   (b) project finance arrangements;
   (c) environmental impact analysis/assessment;
   (d) insurance policies (see Clause 31 (Insurance));
   (e) statutory inspections and licences;
   (f) statutory application and permit to discharge effluent;
   (g) control and recording of waste disposal (see Schedule 5 (Environmental protection and waste disposal));
   (h) planning permission;
   (i) permits required for use of the Site (see Sub-clause 7.1).

Guidance on compiling Schedule 2: Documentation

See Clauses 20 (Confidentiality) and 21 (Documentation)

A disciplined approach should be established to the generation and issue of Documentation particularly the list of Documentation to be produced or approved as the project proceeds.

Documentation for approval

The Project Manager will almost certainly wish to approve certain key drawings or other documents. The procedure for submitting and approving them is laid down in Clause 21. In the case of a reimbursable contract the design may not initially be sufficiently well defined to enable a comprehensive list of drawings to be established. If this is the case an attempt should be made to develop a generic list in this Schedule of documents that are to be subject to this procedure under the heading 'Documentation for Approval'. While recommendations can be made that are not always applicable, the following may be considered for inclusion in the list:

(a) drawings or other documentation required by the Project Manager for local and other statutory authorities;
(b) process block flow diagrams including mass and energy flows and quality relating to battery limit conditions;
(c) process control and instrumentation diagrams;
(d) key process equipment drawings and specifications;
(e) layouts (plot plans and elevations);
(f) drawings locating battery limit crossings;
(g) electrical single line diagrams;
(h) safety plan.

Occasionally important design calculations, specific design details of critical equipment items or other technical documents may be considered to require the Project Manager's approval.

Whenever possible the essential Documentation should be discussed and agreed during the tender negotiations and incorporated into the Specification, so that it becomes part of the Contract and needs no further approval. In a reimbursable contract the Documentation to be approved later is likely to be significant in volume.

Sub-clause 21.6 also deals with documents not falling within the scope of Schedule 2. If certain categories of document are to be excluded from the Project Manager's rights of inspection, they should be described in this Schedule or in a Special Condition.

Documentation for information only

If the Project Manager wishes to see certain types of document under the terms of Sub-clause 21.7 as a matter of course, these may be listed in this Schedule under the above heading, thus avoiding the need for the Project Manager to give notice on each occasion. The list should include the type of document, the subject to which it refers, the stage of its completion and the form in which it should be submitted.

Final Documentation and Manuals

These are referred to in Sub-clause 21.8 and are likely to include drawings necessary for the Purchaser's operation and maintenance of the Plant, mechanical manuals, spares lists, start-up and operating instructions and similar documents.

Final Documentation should include but is not limited to:

(a) Safety File (a requirement of the CDM Regulations);
(b) drawings;
(c) general arrangements;

(d) details;
(e) manufacturer's drawings and manuals;
(f) documents;
(g) process data sheet;
(h) mechanical data sheet;
(i) specification for painting, insulation, marking;
(j) designers' calculations, including data processing output;
(k) material test certificates;
(l) pressure test and lifting gear certificates;
(m) inspection reports;
(n) operating instructions;
(o) lubrication instructions;
(p) maintenance instructions;
(q) list of replaceable parts, with specifications and expected working life;
(r) list of recommended spare parts with specifications;
(s) CAD generated data in the form of discs, tapes, etc.;
(t) statutory records:
    (i) planning approval;
    (ii) building permits;
    (iii) health and safety records;
    (iv) HAZOP studies;
    (v) environmental impact studies;
    (vi) method statements or descriptions of how the plant is to be constructed;
    (vii) pressure test certificates and other forms of non-destructive testing;
    (viii) design documentation relating to pressure systems;
    (ix) fire permits;
    (x) structural calculations;
    (xi) validation data and protocols.

It should be noted that the above list is not exhaustive and the Contractor should confirm the full extent of the statutory documentation required by the national and local authorities having power over the site and any new plant to be built thereon.

The Schedule should also state the number of copies to be provided.

Purchasers should bear in mind that a fully comprehensive set of final, as-constructed drawings could prove very expensive. Purchasers should therefore consider specifying only those as-constructed drawings and documents which they would regard as essential. If the Purchaser does not require these to be provided by the Contractor, this should be expressly stated.

Final Documentation may also include validation and qualification documents all duly signed and approved by appropriate personnel from the Purchaser and Contractor.

The Purchaser should review, after perhaps six months of commercial operation, how the Plant is actually being operated and whether or not it complies with the designer's intention as regards both safety and performance, and whether or not the manuals and drawings need modification in the light of practice. It is also useful for the Contractor to revisit the Plant periodically to gain and collate practical feedback on Plant operation, for the benefit of future design (subject to any constraints of commercial confidentiality).

Process design manual

It is good practice to prepare an operating manual describing how the process is designed to work, how to start-up the Plant and shut it down safely, and what should be monitored while it is running to achieve the maximum available efficiency. It is also vital to the safe operation of the plant that the Purchaser, who is normally responsible for operating activities, is fully aware of the detailed parameters and operating principles which are the basis of the design.

If, after the Plant has been running for some time, a change takes place in some external factor such as the feedstock composition, it may be necessary to review the process design in order to decide on new operating conditions which will accommodate the change and still optimise the Plant performance. Alternatively, a deterioration in performance may occur without any obvious cause, or, again, the Purchaser may call for a review of capacity. The resulting 'troubleshooting' and 'debottlenecking' exercises will require the Purchaser to know more about the original Specification than simply the final equipment data; he will need to know what compromises were made between minimum capital cost and minimum operating cost, what safety margins were used in arriving at sizes or thicknesses of materials subject to stress or corrosion, what additional heat transfer surface became available by virtue of the use of standard tubing larger than the minimum theoretical size, and so on.

It is therefore recommended that in addition to the operating, maintenance and safety manuals, a process design manual should be included in the list of final documentation and manuals. The process design manual should include the background information on process design, and particularly state how compromise decisions have been made. A similar consideration applies to computer software. In the event that changes are made to the plant design or to its as-built state, these changes should be clearly recorded and then included in the final issue of the process design manual.

Guidance on compiling Schedule 3: Responsibilities of Purchaser

See Clause 4 (Purchaser's responsibilities) and Guidance on compiling Schedule 1 (Description of the Works)

Unless the Contractor is to provide the process on a licensed basis, it is common for the process design of the Plant to be provided by the Purchaser. In the case of reimbursable contracts covered by the Green Book, it is rare for the Purchaser not to be closely involved in the design, engineering and construction aspects of the Plant, frequently by way of an integrated team of the Purchaser's and the Contractor's personnel. This 'integrated taskforce' is usually located in the Contractor's offices. This Schedule should list the activities for which the Purchaser will be responsible, and the times at which these should be completed so that the Contractor's progress is not delayed. It is important that the Purchaser appreciates that he has an obligation to provide specific items, particularly information, for the Contractor and that timely performance of this obligation is essential. It should be noted that although Schedule 1 may contain outline information about any design and engineering to be carried out by or on behalf of the Purchaser, the details must be set out in this Schedule.

In most cases the Purchaser will provide information in advance of the Contractor starting work. Some of this may be contained in the Specification and its attachments, but all the Documentation to be provided after the date of the Agreement should be described in as much detail as possible in this Schedule.

The list of activities provided at the end of Schedule 1 may be used as an *aide-mémoire* in checking that all the relevant activities to be undertaken by the Purchaser (other than those specifically mentioned in the Conditions) are included in this Schedule.

If technology is licensed by the Purchaser from a third party, this Schedule should define the role of the licensor's personnel in checking or advising on design work, specifications, construction, start-up and performance testing, as applicable.

In some instances, the Purchaser may elect to procure certain items of equipment and 'free-issue' them to the Contractor. The 'free-issue' items may, for example, be special process equipment, which the Purchaser wishes the Contractor to install, or 'long lead' equipment items that have to be ordered by the Purchaser before he has engaged the services of the Contractor. The items must be clearly identified and delivery dates specified.

In Clause 23 (The Site) it is stated that the Purchaser will give possession of the Site to the Contractor, and access to it. This may involve the construction of a roadway, gates, fencing, etc., and the extent to which these will be provided by the Purchaser should be specified in this Schedule.

Where the Site is part of an existing facility of the Purchaser, with access controlled by his security staff, the Purchaser will need to issue instructions regarding the availability of permits for individuals and vehicles to enter or leave the premises, and also about observing any restrictions on access to other parts of the facility. The latter should be clearly defined by suitable signs. Note also the requirements relating to the CDM Regulations referred to in the Guidance on compiling Schedule 4 (Health and Safety).

Copies of any rules applying to personnel on the Site must be supplied by the Purchaser to the Contractor. If compliance with any of these could increase the Contractor's costs, they should form part of the enquiry against which the Contractor is invited to quote. Such rules could affect safety, working methods or industrial relations.

The Purchaser may choose to provide security, welfare and other facilities for use by the Contractor. If so, these should be detailed in this Schedule.

The extent of the Purchaser's provision of facilities for Health and Safety should be in this Schedule, as should the extent to which the Purchaser will be responsible for the disposal of products and solid or liquid wastes arising from the operation of the Plant, and the circumstances under which this will apply (see Guidance on compiling Schedule 5 (Environmental protection and waste disposal)).

Guidance on compiling Schedule 4: Health and Safety

See Sub-clause 7.2 and Clause 26 (Health, Safety and Environment)

In the United Kingdom, the Health and Safety at Work, etc. Act 1974 (as amended by Schedule 3 to the Consumer Protection Act 1987) imposes obligations on the Purchaser, the Contractor, the Subcontractors and others to ensure that the Plant is safe to construct, operate and maintain and that adequate documentation and other information is provided to ensure that the Plant can be operated and maintained safely and without risks to health. It is not possible to transfer either the obligations or the penalties for failure to meet them. In the case of a process plant, however, there is room for genuinely held differences of opinion as to what is or is not 'reasonably practicable' or 'safe' or 'without risk to health'. It is essential that neither the Purchaser nor the Project Manager can impose contractually upon the Contractor a requirement to do something that the Contractor considers to be a risk to health or safety.

In drawing up the Specification and Schedule 1 (Description of the Works), both parties should satisfy themselves that the health and safety risks are fully assessed and eliminated so far as reasonably practicable. The parties should remember that in the event of any incident the Purchaser, the Contractor and the designer if a separate organisation may be liable to prosecution under the Act.

It should be noted that Clauses 3 (Contractor's responsibilities), 16 (Variations) and 17 (Contractor's Variations) are particularly relevant where any unforeseen hazard comes to light in mid-project. Risks to health or safety are hazards and should be treated as such. Rather than run any risk of borderline areas being neglected, overlapping of zones of responsibility of the Contractor and the Purchaser for checking designs and the quality and completeness of work should be encouraged.

Preparation of the Safety File is an important element of the work of the Project. This is referred to in the Guidance on compiling Schedule 2 (Documentation) as part of the Documentation to be passed to the Purchaser by the Contractor.

The Health and Safety Executive publishes *Guidelines on the Preparation and Operation of Contracts in the Petroleum Industry* (1987), which provide guidance to contractors and purchasers on their responsibilities for health and safety. Although the advice relates to the oil industry, most points are also of relevance to the process industry in general.

All relevant United Kingdom legislation should be taken into account by the parties; for example, the Management (Health and Safety) Regulations, the CDM Regulations, the Control of Major Accident Hazard Regulations (COMAH). Specific appointments must be made by the Purchaser in order to comply with the CDM Regulations.

It is to be expected that the Contractor will act as Principal Contractor under the CDM Regulations. If this is not to be the case it should be clearly stated in this Schedule and reference should be made to the organisation appointed to this role.

Guidance on compiling Schedule 5: Environmental protection and waste disposal

See Clauses 26 (Health, Safety and Environment) and 34 (Site clearance)

The impact of environmental protection on the design and operation of manufacturing plants has become highly significant. As a result, all parties involved in the contract should be aware of any national and international laws and treaties which affect the project.

For those projects which could have an impact on the environment, it is the duty of both the Purchaser and the Contractor to establish a properly-considered analysis of the impact on the environment of the plant, its products, by-products and wastes. Whilst some issues are relatively simple, requiring disclosure of the level of release to the environment or statements regarding volumes, analysis and frequency of release, others are significantly more complex. In the latter case a detailed Environmental Impact Assessment is usually required by environmental protection agencies, and this should be undertaken by a qualified specialist employed or retained by the Purchaser.

In the United Kingdom, and some other parts of the world, the ultimate responsibility for pollution usually lies with the operator of a site or plant. This is particularly true regarding issues such as the following:

(a) disposal of waste;
(b) contamination of air, sea, land and ground water;
(c) disposal of packaging materials or product containers;
(d) the handling and disposal of nuclear and radioactive chemicals.

The latter need special consideration due to the potentially long period of nuclear activity of the chemicals handled. It should be noted that in Europe and most industrial countries it is a statutory duty to consult the appropriate Government Inspectorate regarding the handling, storage and disposal of radioactive material. Specialist advice should be sought in the event that either the Contractor or the Purchaser are not sufficiently experienced in the handling and processing of such materials.

In addition, many countries now expect operating companies to use recycled building materials in construction wherever possible to minimise the use of non-renewable or scarce resources.

A further issue to consider is that of any tax applicable to landfill, particularly in respect of contaminated materials. In some jurisdictions it is permissible to apply for exemption from such landfill taxes in prescribed situations. It can be the case that anyone connected with the work is permitted to make such an application. Clarity is therefore necessary in considering who in the contractual chain is best placed to make such an application and how the resulting savings in tax payments will feed into the contract payment mechanism. It may for example be appropriate that a relevant subcontractor obtains the tax exemption (which generally is otherwise payable on delivery of the waste to the licensed landfill site), and allows an agreed percentage of the savings in his sub-contract price. If the obtaining of such exemption is time-consuming it would be appropriate that he is rewarded for his additional work by not having to pass all the saving back to the Contractor and hence to the Purchaser.

Clause 34 clearly places the responsibility for Site clearance with the Contractor. However, it must be recognised that there are a variety of types of waste other than those specifically stated which may require disposal. Typical forms of waste may be as follows:

(a) Standard non-hazardous waste resulting from the construction of the Plant. The removal of this waste is generally the responsibility of the Contractor, but under the reimbursable form of contract any costs arising from the removal would normally lie with the Purchaser. This matter should be clarified prior to contract signature.
(b) Waste resulting from Site preparation or demolition which may be a mixture of hazardous and non-hazardous materials. Asbestos or toxic wastes are typical of the hazardous materials which may be found on an existing chemical manufacturing site. The Purchaser normally bears the responsibility of ensuring that such hazardous materials are removed from Site safely and disposed of at a designated or licensed site. The Purchaser may request the Contractor to organise the necessary specialists to carry out this work and it is expected that in so doing he will operate all necessary safety procedures to comply with the Codes of Practice and legislation which apply. Nevertheless the ultimate responsibility still lies with the Purchaser and he must oversee each stage of the demolition or Site clearance where hazardous materials are being handled, to satisfy himself that he is discharging his legal obligations in this regard.
(c) The final main category of waste product requiring clearing from Site is chemical substances which are hazardous, toxic or possibly of unknown character resulting from chemical processing in the Plant. In this instance it is normally the sole responsibility of the Purchaser to arrange for the safe removal and disposal of the wastes. An exception to these circumstances would be if the Contractor is the owner of the process technology, in which case he may be best placed to establish the correct protocols for the removal from Site of any waste, by-product or generally hazardous materials from his process, but the actual removal and disposal of these will be the responsibility of the Purchaser.

Guidance on compiling Schedule 6: Quality assurance and validation

See Sub-clause 3.8

In broad terms, quality assurance requires a documented systematic approach to the attainment of quality.

Quality assurance is now widely accepted internationally as mandatory in the manufacturing and construction industries. In British Standard BS 4778, 'Quality Assurance' is defined as 'All activities and functions concerned with the attainment of quality' and 'Quality' is defined as 'The totality of features and characteristics of a product or service that bear on its ability to satisfy a given need'.

There are several internationally-recognised quality assurance standards, but ISO9001: 2000 'Quality Systems: Part I Specification for design/development, production, installation and servicing' is particularly applicable to the process industry. A useful guide to that Standard, outlining its application to a process plant project, has been published by IChemE, 'Quality Assurance: A Guide to the Application of ISO9001 to Process Plant Projects', available from IChemE Book Sales.

In addition to the standards for quality, specific engineering or other standards are likely to be listed in the Specification. Care must be taken to define to what extent each party is responsible for producing record documents.

The Purchaser must advise the tenderers at the pre-contract stage of any specific quality assurance issues that are to apply to the project, and he should provide copies of quality assurance control procedures that the Contractor will be required to adopt; for example:

> 'The Contractor shall compile indexed and cross-referenced records comprising such Documentation as drawings, weld procedures, x-ray images, and material and test certification. Such records shall be subject to review and approval by the Project Manager in accordance with Sub-clause 21.2 of the General Conditions.'

Before compiling Schedule 6, detailed discussions should take place between the Purchaser and the Contractor regarding the extent and complexity of the quality assurance activities which will be required during the project. As a minimum, the contents of a master plan should be included in the Schedule, with a tabulation of the responsibilities of the Purchaser, Contractor, Subcontractors and Supplier clearly allocated. The Schedule should also state whether third party inspectors or quality assurance specialists are to be involved and what responsibilities they may have for approving documentation. If the quality assurance activities are critical for completion, the times and duration of such activities should be included in the Approved Programme.

It should be recognised that it may not be possible to define the exact scope of the quality assurance work to be carried out by each party, because at the time of signature of the Agreement the detailed design will not be available. In spite of this situation occurring on most occasions, every effort should be made to define the scope of work anticipated so that the costs to be borne by the Purchaser are reasonably predictable and within his budget for the project.

The following schedules may refer to quality assurance activities, and they may require elaboration as the project proceeds:

Schedule 2 (Documentation);
Schedule 13 (Pre-installation tests and procedures);
Schedule 15 (Take-over procedures);
Schedule 16 (Performance tests and procedures).

When an event such as payment or taking over is dependent upon quality assurance (or validation) activities, this should be clearly and unambiguously cross-referenced in the appropriate schedule.

Validation may be required as part of the overall quality assurance programme. The purpose of validation is to create documentary evidence providing assurance that the product from a manufacturing process will consistently meet its defined specifications and quality characteristics in all required circumstances. In particular the pharmaceutical, electronics and aerospace industries apply rigorous techniques of this nature, and if the Purchaser or the Contractor is unfamiliar with such activities, they should engage expert advice before signing a contract in which validation is included.

If it is to be applied, validation needs to start at the development phase (that is, pre-contract) by defining the product specifications in all respects, including the deviations which can be tolerated, but which must never be exceeded when the plant is in operation. The principal purpose is to ensure that the plant meets the design intent in every respect.

The quality assurance Master Plan should include sections on qualification activities which, when they have been completed, will provide detailed documentation on all the checks that have been carried out during design, procurement, fabrication, installation and testing to ensure the robustness and reproducibility of the process when in operation.

Guidance on compiling Schedule 7: Subcontracting

See Clauses 9 (Assignment and subcontracting) and 10 (Nominated Subcontractors)

In process plant projects most, if not all, of the equipment and other materials will be manufactured and supplied by a number of Subcontractors. Similarly, some or all of the construction work on Site will normally be subcontracted.

A Purchaser may wish to ensure that the design of a particular process unit or plant item is the responsibility of designers with appropriate specialist experience, and in that case the Purchaser should make this clear at the tender enquiry stage. A contractor who is tendering for a project may himself have some specialised design experience, but may need to subcontract other parts of the design to specialist subcontractors, in which case he should make this clear in his technical tender.

If the Purchaser has stated in the tender enquiry that he requires certain design to be carried out by named specialists, they should be nominated by the Purchaser in this Schedule.

A Purchaser who is subject to European Union Procurement Directives (and the national law implementing them) or similar directives in other countries will no doubt bear in mind this legislation if he intends to nominate particular Subcontractors, provide an approved list, or in any other way restrict the Contractor's free choice of Subcontractors. Such matters should be addressed in this Schedule.

If the Contractor has stated in his tender that he proposes to subcontract parts of the Works to named specialists, and the Purchaser accepts them, this should be added to this Schedule before contract award.

In reimbursable contracts, because the Purchaser bears the financial burden of the payment to each Subcontractor, he is directly concerned with the selection of the Subcontractors and has the right to make selections of his choice. The Contractor should therefore provide the Project Manager with sufficient information (including a recommendation, if possible) to enable him to make his choice on behalf of the Purchaser. This Clause provides the framework for a subcontracting procedure designed to safeguard the interests of both parties.

Sometimes the Contractor has links with other companies which may have an interest in obtaining subcontracts. The purpose of Sub-clause 9.3 and 9.5 is to ensure that the Contractor does not abuse his position in this respect.

The Conditions assume that the Contractor will be contracting in his own name and that payments will flow from the Purchaser to the Contractor, who in turn will pay the Subcontractor.

A detailed procurement procedure should form part of this Schedule including, if applicable, provision for payment by the Purchaser into an account to be operated by the Contractor for payment of Subcontractors as mentioned in the Guidance on compiling Schedule 19 (Terms of Payment).

Clause 10 deals with nominated Subcontractors and in particular specifies the procedure to be adopted if the Contractor has valid reasons for objecting to any Subcontractor nominations which the Project Manager has made.

IChemE publishes Forms of Subcontract suitable for use with these conditions (the Yellow Book and, specifically for civil engineering works, the Brown Book, available from IChemE Book Sales).

Guidance on compiling Schedule 8: Contractor's named personnel

See Clause 12 (Contract Manager and Contractor's staff)

Many Purchasers, with some justification, regard the experience of a particular Contract Manager to be of prime importance, and so there is a provision in Sub-clause 12.1 that the Contract Manager will not be replaced without the consent of the Project Manager.

Where specialised experience is required to carry out a project, the Purchaser may consider it prudent or, in some cases, essential to request the Contractor to nominate not only the Contract Manager but also key engineering and/or construction supervisory staff. Sub-clause 12.4 provides for the continuing participation in the project of named key personnel who may be of considerable importance to the success of the project. These nominations should be agreed pre-Contract and a list of such named personnel will be included in this Schedule.

Whilst Sub-clause 12.4 recognises that the Contractor may wish to replace key personnel identified in the Contract, such replacement must be with the written consent of the Project Manager. If it is considered necessary, this Schedule should specify the circumstances under which the Contractor may nominate a replacement and any required overlap period for handing over to the replacement. The Project Manager will be expected not to withhold his consent unreasonably.

Guidance on compiling Schedule 9: Training by Contractor

See Sub-clause 3.9

When the Contractor has an obligation under the Contract to provide training for the Purchaser's staff, the Purchaser has the responsibility to make personnel of suitable calibre available for training. The most satisfactory arrangement for practical training to be achieved is for the Purchaser to appoint and make available the key operating and maintenance personnel before completion of construction and well before commissioning so that they can work alongside the Contractor's engineering, construction and commissioning personnel. If, for example, they can assist the construction personnel in making lists of outstanding minor items requiring modification or finishing, this will familiarise them with the equipment and control systems, particularly pipework and the position of valves, vents and drains.

Training covers a broad field and the Contractor may be called upon to provide basic training on various aspects of plant operation, health and safety, quality assurance, instrumentation, control systems and maintenance procedures. Where the plant includes technology with which the Purchaser's personnel are not familiar, induction courses and workshops should be arranged explaining the process, philosophy of design and control systems and start-up, shutdown and emergency procedures.

All these requirements should be listed in the Schedule, together with the methods and procedures to be adopted in carrying out the training, the personnel who are to be trained and the timing and duration of the training.

Guidance on compiling Schedule 10: Parts with limited working life

See Clause 37 (Liability for Defects) and Guide Note M (Liability for Defects)

Leaving aside normal wear and tear, it may be economical to specify certain replaceable parts, such as filter cloths, with a short working life because the cost of a more durable alternative would be many times higher. Such parts should be detailed in this Schedule. A suitable introduction to a list of such parts could be:

'The following parts have been specified by agreement to have a limited working life, on economic grounds. Attention is drawn to Sub-clause 37.12 of the General Conditions in respect of such parts.'

The list should identify the item of equipment in which the part is fitted, a description of the part including any identification code, the estimated working life of the part and the guaranteed minimum working life under normal running conditions.

Guidance on compiling Schedule 11: Times of completion

See Clause 13 (Times of completion and Approved Programme) and Guide Note S (Completion, taking over, testing and start-up)

The Contractor's work may have a single objective of significance to the Purchaser in terms of timeliness of completion, such as the construction and start-up of a process unit; alternatively it may have several such objectives of varying priority. These will in either case form the main features of this Schedule and hence the Approved Programme, and should be considered at an early stage of the project.

Except in the most simple of projects, the Plant is likely to consist of several sections, some of which are operationally dependent on others. There will thus be a logical sequence in which completion of each will be required in order to achieve the most satisfactory commissioning of the whole project. Furthermore the actual task of thoroughly inspecting and testing a completed plant before it can be used safely can be costly in time and resources. To leave the whole of a project to be inspected and tested at the same time could lead to the procedure being rushed because of pressure from production management and hence lead to potential hazards, or else lead to unexpected delay if the inspection work is done properly. For this reason adequate time should be allowed for inspection and testing.

For each event for which the Purchaser requires to define the time-scale, this Schedule should specify:

(a) the time for commencing the event (either as a date, or by a period from the date for the commencement of the Works as defined in the Agreement or some other defined event); and
(b) the time for completing the event (by date or period).

Appropriate cross-references should be made to the relevant Schedules, in particular Schedules 14 (Criteria for the completion of construction), 15 (Take-over procedures) and 16 (Performance tests and procedures).

Where several sections of the Plant or the Works are identified in the Specification or in Schedule 1 (Description of the Works) separate dates or periods should be stated for each.

In any event clarity is essential as to what is intended as an event for which a date or period is stated.

The Schedule should only contain events connected with liabilities for delay and other key events which are of importance to the Purchaser, or which are required for the purposes of Sub-clause 13.2 (see below). Care should be taken not to unnecessarily restrict the Contractor's freedom to programme the work as he sees best, and thereby add to the cost.

Sub-clause 13.2 states that the Project Manager shall certify the occurrence of any event which is identified in Schedule 11 (Times of completion) as being an event which triggers a payment by one party to the other—usually from the Purchaser to the Contractor. Any such event must therefore be clearly defined in Schedule 11, with no ambiguity as to what exactly the event is to be. Such events typically relate to lump sum items (for example, construction of an access road) or the delivery of a major item of plant. Any such lump sum will be a matter for Schedule 18 (Cost elements, rates and charges).

To avoid as far as possible the operation of Sub-clause 14.2, where the Contractor may become entitled to an extension of time, the events listed in this Schedule should be under the control of the Contractor without his having to rely on the performance of the Purchaser or the Project Manager. For this reason events such as the 'Plant coming on stream' or the 'issue of an Acceptance Certificate' are not appropriate for the Contractor to have to achieve to avoid paying liquidated damages. Furthermore undue haste during the initial period of operation of a process may mean taking avoidable risks with the safety of the plant and the personnel.

Sub-clauses 13.1 and 15.1 are worded in a way that places emphasis on the completion of construction as the event most appropriate for the application of liquidated damages for delay, but the Purchaser may prefer to apply such damages to a delay in taking over. If that is the case, the Purchaser should extend the list of criteria in Schedule 14 (Criteria for the completion of construction) to include all the procedures he wants the Contractor to carry out ready for taking over by the Purchaser (see the check list in Guide Note S), and then write 'None' in Schedule 15 (Take-over procedures). Then the Project Manager could issue a Take-Over Certificate under the terms of Sub-clause 33.7.

Such an arrangement would avoid any confusion about how far take-over procedures had been completed by a certain date, which could arise as a result of the application of the terms of Sub-clauses 33.8 to 33.12 inclusive.

Whatever the event, it requires precise definition to minimise the possibility of dispute as to the time when it is actually achieved.

Guidance on compiling Schedule 12: Liquidated damages for delay

See Clauses 4 (Purchaser's responsibilities) and 15 (Damages for delay) and Guide Note S (Completion, taking over, testing and start-up)

The Schedule must clearly define the amount of liquidated damages and the events to which they apply.

The events should correspond with those listed in Schedule 11 (Times of completion), and the amounts should be no more than a genuine pre-estimate of the loss which would be suffered by the Purchaser if the Contractor was late. English Law will not recognise penal damages, and the imposition of excessive amounts would be unenforceable.

It is common practice in cost reimbursable contracts for liquidated damages for delay not to be levied, or at least the maximum that might be levied to be more strictly limited than on a lump sum contract.

If the Purchaser is willing to limit the Contractor's liability for delay (which could be reflected in lower Contractor's rates or fees), an upper limit for the total liquidated damages payable for delay could be stated, either as a percentage of the Contract Price or as a sum of money (which would equate to a multiple of the daily/weekly rate of damages). Purchasers should note that, in accordance with Clause 15 the Contractor has no liability to pay further damages once an upper limit is reached.

If liquidated damages for delay are not to apply at all, this must be stated explicitly either in this Schedule or in a Special Condition. Remember that if this Schedule states that liquidated damages are 'nil' or 'zero', the Court has interpreted this as meaning that liquidated damages do apply but are of no financial effect and that no unliquidated damages may be claimed either.

An alternative approach to liquidated damages for delay, with their necessarily negative approach of imposing an addition financial burden on the Contractor, is for the Purchaser to pay a bonus for timely or early completion (see Guide Note H (Incentives for early completion or improved performance).

Guidance on compiling Schedule 13: Pre-installation tests and procedures

See Clause 22 (Inspection and pre-installation tests)

Clause 22 provides for the Project Manager to carry out off-site inspection and tests to satisfy himself (or his nominee—see Sub-clause 22.8) that the Materials are in accordance with the Specification.

Such Materials could include machines or package units the performance of which the Purchaser requires to be demonstrated in the manufacturer's premises by a running test. A suitably worded schedule should be included in the Contract.

Pre-installation and factory tests could include not only tests of the performance of components, but any appropriate forms of test, such as pressure testing or load testing. The inspections could cover welding records, certificates of origin of alloys, etc.

This Schedule should include:

(a) a description of the work or Materials to be tested;
(b) a description of the test and, where appropriate, details of the test rig and instrument accuracy;
(c) the criteria to be achieved during each test together with acceptable tolerances; and
(d) the action to be taken by the Contractor if the work or Materials fail to pass any test after the Contractor has had an opportunity to rectify or replace them.

Guidance on compiling Schedule 14: Criteria for the completion of construction

See Clause 32 (Completion of construction) and Guide Note S (Completion, taking over, testing and start-up)

This Schedule will define the criteria to be satisfied by the Contractor when he offers the many elements of the Plant for inspection and provides the Project Manager with a draft form of Construction Completion Certificate.

This Schedule should reiterate that the Plant must be constructed in accordance with the Specification. It should then list the activities and tests which the Purchaser requires to be carried out by the Contractor within the time limit stated in Schedule 11 (Times of completion), such as those marked (a) to (y) in the list in Guide Note S.

Guidance on compiling Schedule 15: Take-over procedures

See Clause 33 (Taking over) and Guide Note S (Completion, taking over, testing and start-up)

This Schedule should include all the inspection procedures, equipment tests, take-over tests and other plant checks and documentation provided by the Contractor and needed to satisfy the Purchaser that his operating staff can safely start-up, operate, shutdown and maintain the Plant except those already included in Schedule 14 (Criteria for the completion of construction). The parties to the Contract may decide at the time of signing that this Schedule cannot be satisfactorily completed until the Contractor has had time to carry out further design and engineering work, in which case a Special Condition should be included stating the time by which a draft of the complete Schedule should be submitted to the Purchaser by the Contractor.

It will be a matter of judgment for the Purchaser to decide how much of the work involved in finalising the preparations for start-up should form part of the take-over procedures, when the Contractor is still in charge of the Plant, and how much should be done after the Purchaser has taken over the Plant.

Referring to the check-list in Guide Note S, the task of charging catalysts could be done by the Contractor, with the assistance of the Purchaser's staff, but charging raw materials, process chemicals and fuel should be done by the Purchaser after taking over, since he is responsible for supplying them.

Guidance on compiling Schedule 16: Performance tests and procedures

See Clause 35 (Performance tests) and 36 (Acceptance), Guidance on compiling Schedule 17 (Performance guarantees and damages for failure) and Guide Note S (Completion, taking over, testing and start-up)

Clause 35 is to apply to the Contract if the Contractor has made specific guarantees in respect of the performance of the Plant verifiable by performance tests (see Sub-clause 35.1).

Guarantees of performance to be confirmed by tests provide an additional opportunity for the Contractor to be in breach of contract, if in the event the guaranteed level of performance is not achieved. The damages consequent upon such a breach may be established in advance as liquidated damages, or left to be quantified after the event as unliquidated damages. The IChemE Red Book approach, reflecting normal industry practice, is to allow for a defined margin of failure of the guarantee to be covered by liquidated damages. Failure outside the defined margin is a matter for negotiation. One approach is to agree a reduction in the Contract Price. Such an approach sits well with a lump sum contract where the contractor has a full opportunity to price the risk of failure of the guarantee.

In a cost reimbursable contract, as has been noted elsewhere, the Contractor has only very limited opportunities to price for risk. The main opportunity is in his management fee, but even that is normally established on the basis of expected costs of overheads, with an addition for profit, and little allowance for the risk of damages. It is noted in the Guidance on compiling Schedule 12 (Liquidated damages for delay) that it is common practice for such damages not to be levied, or at least for the maximum amount that might be levied to be more strictly limited than on a lump sum contract.

A judgment therefore has to be made as to whether to include performance guarantees and the associated opportunities for damages. That will depend in part on what features of the Plant's performance are under consideration. Performance guarantees which match similar guarantees in a subcontract package are more readily acceptable, as the price-based subcontract will mean that the Subcontractor will be well-placed to price for the risk associated with the guarantee. Performance guarantees which are not so linked are more difficult, unless they relate to a licensed or patented element of the Plant for which the Contractor is the licence or patent owner. In that situation, where the proprietary element of the Plant may well be provided under a lump sum item within the contract, the application of guarantees and the possibility of associated damages do not have the same difficulties, as the lump sum item allows for the pricing of the risk.

Another possible situation is where the Contractor has specific technical or design expertise which the Purchaser relied on when making the appointment of the Contractor. Some form of damages, usually liquidated damages, will probably be sought from the Contractor to cover this aspect of his work. He will have to make an assessment of his risk at the outset based upon his experience in the particular area of expertise.

Certainly if guarantees are to be imposed under a cost reimbursable contract with its philosophy of close working of the Purchaser and the Contractor it is not appropriate to consider a reduction in the Contract Price, and Sub-clause 35.9 differs in that respect from its equivalent in the Red Book in that the Green Book version only allows for liquidated damages.

Careful consideration must therefore be given to the possible inclusion of performance guarantees, with liquidated damages applying in the case of failure, in a cost reimbursable contract.

It may in certain situations be appropriate to require that performance tests which do not directly relate to performance guarantees are undertaken. Clause 35 does not apply to such tests by virtue of the terms of Sub-clause 35.1. However, whether included within the original scope of the Works as stated in the Contract, or ordered as a Variation, there is no reason why the Contractor should not assist the Purchaser with the conduct of such tests. Their contractual significance would depend on what errors or lack of performance they demonstrated, including for example a failure to comply with Sub-clause 3.4 or a Defect requiring correction under Clause 37 (Liability for Defects).

If performance guarantees are to be included, the drafting of Schedule 16 will need considerable care. It is probably the most difficult Schedule to prepare well, and commonly the most neglected. At the time of drawing up the Contract the event seems far away and the details tedious, but careful attention to those details pays dividends by avoiding discord on site at the critical time of performance testing.

The time limit from the date of take-over for the tests to be carried out must be specified here, for the purposes of Sub-clauses 35.9, 35.10 and 35.11.

How the Plant is to be prepared for the tests, the duration of tests, the performance criteria to be measured, conditions governing interruptions, methods of measurement and analysis (which should be standard whenever possible) and tolerances to be allowed to cover lack of precision should all be specified. Routines for measuring stocks at start and finish, recording and interpreting readings on instruments and collecting and preparing samples should all be included. Many of these requirements affect the detailed design of the Plant and so the design office needs to be aware of them. For example, it may be necessary to measure the flow of a stream of solids, when for normal operation an accuracy of say five per cent may be quite sufficient. This might not satisfy a Purchaser if it formed part of a performance test procedure for which batch weighing might be necessary instead of, or as well as, continuous belt weighing.

A plant will usually be required to operate satisfactorily over a range of conditions, and it may be desirable to provide corrections to the guarantees which can be applied if the performance test takes place under conditions differing from the basic design case. For example, the guaranteed capacity of a plant may equal the nominal capacity when the ambient air is at the basic design temperature, but may be reduced by a certain percentage for each degree that the temperature is above the design figure.

Performance criteria to be tested will usually include plant capacity, details of consumption of raw materials and utilities (descriptions, consistency of quality, range of rates of input required, etc.) and essential criteria for the quality of the finished products. In some cases guarantees may also be required for reliability, the consumption of utilities (if these are a significant cost factor) and the quantity and quality of by-products (including steam) and polluting effluents. However, for a performance guarantee to apply, it must be stated explicitly in Schedule 17. It cannot simply be implied or inferred.

This Schedule should also state how the measuring tolerances should be allowed for in calculating the results of the tests for comparison with the guaranteed values. For example, it could have been agreed that the instrument readings will be used without any corrections, so that neither party benefits from any measuring tolerances.

Unspecified performance tests

Sub-clause 35.3 provides for tests to be carried out that have not been specified in this Schedule. If an agreement is reached between the Project Manager and the Contractor that any unspecified tests are to be carried out then they should be treated as Variations. Care must be taken when defining the scope and purpose of any new or unforeseen tests since they may involve an addition to the performance criteria set out in this Schedule. Ideally, additional tests of this nature should be avoided and every effort should be made to define the full extent of the tests required at the pre-contract stage. It should also be noted that such tests might necessitate alterations to the original plant design to enable additional temporary or permanent instrumentation and sample points to be installed. The requirement for both parties to agree that the tests should be carried out is a reasonable method of proceeding, particularly if the timing of a request to carry out such tests would not allow the work to be carried out for reasonable time or cost. The Contractor should be in a position to decline to carry out such tests particularly if he is likely to incur overall delay, unrecoverable costs or additional liability and there should be no implication that such tests should be part of the Contractor's obligations unless required for statutory reasons.

Performance testing not completed by the end of a Defects Liability Period

Finally, Sub-clause 35.13 addresses the possibility of the performance tests not having been completed by the end of the Defects Liability Period. In such circumstances the tests are deemed to have been passed unless the Purchaser can adequately show that the Plant was incapable of passing the tests. This may particularly be appropriate in a situation where a test was failed on one or more earlier occasions and the Contractor has been endeavouring to adjust and improve the Plant so as to achieve the guaranteed performance. Again, the difficulties associated with such guarantees under a cost reimbursable contract are highlighted.

Guidance on compiling Schedule 17: Performance guarantees and damages for failure

See Sub-clause 35.9, Guidance on compiling Schedule 16 (Performance tests and procedures) and Guide Note S (Completion, taking over, testing and start-up)

The appropriateness of including performance guarantees in a cost reimbursable contract is considered in the previous guide note 'Guidance on compiling Schedule 16'.

For each guaranteed parameter (quantity or quality of product or effluent, usage of raw material, utility etc.), this Schedule should include:

(a) guaranteed performance;
(b) maximum permissible deviation; and
(c) the amount of money to be deducted for each unit of failure as liquidated damages.

Such liquidated damages are intended as agreed compensation for the Purchaser's increased operating costs, loss of product, etc., as a result of the Plant's poorer performance, and should be no more than a genuine pre-estimate of the loss which would be suffered by the Purchaser in operating the Plant under such conditions in practice.

Though it is not practicable here to make detailed recommendations, the following may be a useful *aide-mémoire*.

(a) Raw material consumptions, utility consumptions, yields, efficiencies.
These represent direct running costs. Liquidated damages could be set at a figure equalling the extra costs for a certain period of time. Figures between one and three years at flowsheet rate have been used.

(b) Groupings of different running costs elements.
In general, savings in one item should be offset against excess costs of another. Occasionally, however, this is not appropriate—for example, when a particular utility or feed material is in limited supply.

(c) Credit for by-products.
Useful by-products would normally be grouped together with the input materials, the total running costs being arrived at mathematically, but there are instances when a by-product may become a nuisance if made at an increased rate.

(d) Plant capacity.
A true appraisal of the Purchaser's loss through shortage of capacity is very difficult to make, as it is affected by his ability to make full use of the capacity available, but amounts of liquidated damages as a percentage of the Contract Price proportional to the deficiency have often been used.

(e) Product quality.
It is sometimes possible to relate the value of a substance to its purity and to set liquidated damages for running costs. If there were two or more products, their

inter-relation should be considered and credits in the quality of one product set off against shortcomings in another. Maximum permitted limits of impurities may, however, be used to impose restrictions.

(f) Effluent quantity and quality.

Excesses in either may result in higher disposal costs, to which liquidated damages could be related, but absolute limits may be imposed by statutory authorities or other bodies.

(g) Groupings of performance criteria.

It is important to consider whether, and the extent to which, different criteria of performance should be taken together. For instance, a Plant may fall slightly short of the promised output, but be more efficient; another Plant may make a product to the desired specification at appreciable overload but with excessive power consumption.

Performance guarantees should only apply to the really essential parameters. Damages should be unambiguous with respect to any grouping of criteria and should be defined in a way to make their calculation simple.

An upper declared limit of total liquidated damages relating to Plant performance may be set. In some instances, it may be appropriate to set the amount of liquidated damages at zero.

The parties should note that if the agreed contractual position is that the Contractor is to be liable for damages at large, or unliquidated damages, an explicit statement to such effect should be included in a Special Condition. If the damages are to be set at zero, this needs to be stated in Schedule 17. This should also have been made clear in the invitation to tender.

It should also be noted that Clause 35 (Performance tests) does not prescribe the action to be taken in the event that the result of a performance test lies outside a maximum permissible deviation. Bearing in mind that in a reimbursable contract the Contractor is not constrained by a need to design down to a price, any shortfall in plant performance may be due to a design error covered by the terms of Sub-clause 37.3 (b), which places the liability to pay for the cost of rectification on the Contractor, which in turn would be limited by Clause 44 (Limitation of liability) in the General Conditions.

Guidance on compiling Schedule 18: Cost elements, rates and charges

See Clauses 18 (Contract Price) and 40 (Contingencies and prime cost sums) and Guide Note D (Unforeseen ground conditions)

The concept of a cost reimbursable contract is just that—to reimburse the Contractor his costs. Those costs arise from the use of resources by the Contractor in carrying out the Works and delivering the Plant.

Any project, irrespective of size or nature, can be sub-divided into main cost groups under which all costs can be allocated, such as:

(a) head office costs (including contract management, design, engineering, procurement, expediting, inspection, construction, commissioning, general administration and business support facilities generally);
(b) services subcontracts;
(c) materials and freight (materials, equipment, freight, insurance and import duties);
(d) site costs (site office, construction, labour and subcontracts, construction costs including tools and equipment, commissioning).

Each activity involving staff may be further sub-divided for charging purposes to take account of man-hours booked, payroll burden, accommodation, overheads, expenses and other related costs.

The payroll burden element can be expanded to indicate the main items that are usually included, e.g. statutory holidays and vacation pay, luncheon vouchers, national health insurance, pension contributions, sickness pay, employers' portion of pension and life assurance schemes, training levies, etc.

Accommodation and fixed overheads would normally include rates, rents, light, heating, maintenance, building services and security.

General Administration covers such items as legal, accounting, marketing, R&D, general consultancy fees, insurances, etc.

Miscellaneous administrative facilities would include communication costs, stationery, reprographic charges, fees and expenses incurred in the execution of the contract, etc.

Construction labour should identify the known bases for payment, for example, wages, incentives, overtime and premium pay, vacation pay, severance pay, radius and lodging allowances, travelling time and other wage elements introduced by statutory regulation and/or union agreement for all weekly or hourly paid site personnel, statutory payroll taxes, pension contributions, etc.

Construction costs must account for all miscellaneous charges relating to the construction including temporary facilities, power, water and fuel supplies, security, recruitment, transportation, small tools and consumables, etc.

The following list of main cost elements is given as an indication of the various costs, and should not be regarded as an exhaustive list.

Typical list of main cost elements:

(1) Head Office:

(a) Project-related costs:
- (i) salaries—net cost (permanent or agency);
- (ii) payroll burden;
- (iii) payroll accounting;
- (iv) accommodation;
- (v) overheads for agency staff;
- (vi) expenses—net cost;
- (vii) insurance;
- (viii) independent consultants or inspectors;
- (ix) accounting for specified site personnel;
- (x) computer time;
- (xi) communications;
- (xii) licence fees and know-how;
- (xiii) stationery;
- (xiv) prints, reprographic services;
- (xv) provisions for Purchaser's staff.

(b) General overheads:
- (i) marketing, sales and advertising;
- (ii) proposals;
- (iii) accounts in general;
- (iv) laboratory work;
- (v) legal;
- (vi) general administration and overheads;
- (vii) financial items—working capital, bank charges, exchange rates, VAT (if not recoverable), etc.

(2) Construction equipment and tools:
  (a) hired construction equipment (rent and maintenance);
  (b) contractor-owned construction equipment;
  (c) small tools;
  (d) consumables;
  (e) oils, fuels, and greases;
  (f) transportation;
  (g) licences, duty and tax;
  (h) insurance.

(3) Materials:
  (a) materials as defined in Clause 1 of the General Conditions of Contract;
  (b) spare parts;
  (c) purchasing;
  (d) expediting;
  (e) transportation;
  (f) licences, duty and tax;
  (g) insurance;
  (h) inspection (including by third parties).

(4) Site costs:
  (a) recruitment;
  (b) salaries—net cost (permanent and agency);
  (c) wages, including supplementary payments;
  (d) training levies;
  (e) payroll burden;
  (f) overheads for agency staff;
  (g) expenses;
  (h) safety including induction and certification;
  (i) qualification tests of skilled personnel;
  (j) general overheads and administration;
  (k) buildings and contents;
  (l) power, fuel, water and sanitation;
  (m) communications;
  (n) stationery, reprographic services;
  (o) site services, canteen, medical;
  (p) temporary facilities;
  (q) transportation;
  (r) storage of materials;
  (s) waste management and disposal;
  (t) insurance;
  (u) security.

(5) Site commissioning:
  (a) salaries—net cost (permanent and agency);
  (b) payroll burden;
  (c) overheads for agency staff;
  (d) expenses;
  (e) insurance;
  (f) general administrative overheads;
  (g) vendor's/licensor's/subcontractor's staff;
  (h) feedstock, steam, chemicals, etc.;
  (i) spare parts (commissioning only).

Schedule 18 must set out the arrangements for identifying and paying for these costs, which arise from the use of resources in the Contractor's own organisation and from the use of Subcontractors and suppliers. The Contractor's costs will be reimbursed through one or other of the following mechanisms:

(a) directly chargeable to the Purchaser at net cost;
(b) directly chargeable to the Purchaser at quoted rates;
(c) not directly chargeable to the Purchaser but covered by lump sum payments;
(d) not directly chargeable to the Purchaser but covered by rates charged for other cost elements.

Items which are chargeable at net cost will include all Subcontract work and supplies, for which invoices are submitted to the Contractor.

More difficult to quantify are the costs he incurs from his own organisation—for example, his own construction personnel or Contractor's Equipment. In essence there are two ways of dealing with such costs. The simpler way is to have basic unit cost rates set out in this Schedule 18 covering the various grades of personnel and the various categories of Contractor's Equipment which the Contractor expects to use on the Works. Whether or not those unit cost rates (for example, per hour of work or of use) will include elements for overheads and profit will depend upon whether there are other provisions in the Contract for such items—for some possible approaches, see below. Again the simpler approach is to include such elements in the unit cost rates. However, such an approach is more of a unit price basis of payment rather than a cost reimbursable basis.

An alternative approach is for the Contractor to open his books totally to the Project Manager such that the latter can see the actual cost of employing each individual involved with the Works. This has the advantage that the actual costs incurred will be reimbursed as opposed to a surrogate mechanism which will necessarily have generic categories of people. In providing unit rates for the generic categories, the Contractor (to safeguard his position) will tend to quote rates which are higher than actual cost. He will also have to take account of any rules established in this Schedule 18 as to whether (and when) such unit rates can be reviewed to take account of inflation, pay rises, etc. Payment of the actual costs analysed from the Contractor's books contains no such risk, although proper account must be taken of all the payroll burden costs referred to above.

It is important that there is provision somewhere in the payment mechanisms for profit: the Contractor who is only reimbursed his actual costs (net invoices, actual costs of labour with no facility for profit) will not make any profit and hence the contract is not attractive to his business. With one of the advantages of a cost reimbursable contract being the flexibility as to the amount of work to be done, separating profit from the labour unit rates has the added advantage of avoiding the problems of how much 'mark-up' should be included within individual rates when the requirement for each category of resource is not known at the time of tender.

Care must be taken to ensure that the inclusion of other costs, to cover items for which separate payment is not to be made, is kept to a minimum. In contrast, the use of unit rates facilitates such inclusion of other costs. Such other costs may include, for example, small tools and consumables, and statutory charges (such as training levies).

The basis on which costs are to be reimbursed will therefore have to be set out in Schedule 18, which, if unit cost rates are adopted, will include tables for such unit rates which the Contractor will complete as part of his tender. Matters such as overtime rates, and when the Purchaser is willing to cover overtime costs, need to be stated.

Lump sum items

Many Purchasers see benefits in separating out the costs associated with design and with head office inputs, and paying for these two important items through fixed (lump sum) fees. Others however prefer to deal with these two areas of cost in the same way as site-based costs. Those who opt for a head office overheads fee will generally state that the fee will cover both head office overheads and profit.

Any lump sum fee will need clarity as to what it includes and what if any mechanisms will allow for changes to such fee as the result of Variations ordered under Clause 16 (Variations).

Sub-clause 13.2 provides for events 'the occurrence of which can affect the amounts paid by one party to the other.' This will apply to any payments due under Sub-clause 39.2. Such payments may be lump sum items, which again should be stated in Schedule 18.

The extent to which lump sum prices are incorporated in reimbursable contracts can vary widely. If only a design fee and a management fee (covering head office overheads and profit) are included, there is probably no need for any Special Conditions provided the arrangements are clearly set out in Schedules 18 and 19 (Terms of payment).

Where it is proposed to cover the provision of certain Materials and/or Works by one or more lump sum payment, it may be appropriate to consider whether the contract provisions are wholly appropriate and whether any Special Conditions are necessary.

Key issues (other than the inclusion in the Specification and Schedule 1 (Description of the Works) of a fully detailed specification and scope of all the Materials and Works included in any lump sum element) to consider are:

(a) clarity as to the respective responsibilities of the parties (i.e. who does what and when);
(b) liabilities for Defects (see also Guide Note M (Liability for Defects)).

Generally price-based contracts include for the contractor bearing the cost of Defect rectification (see for example the IChemE Red Book). Under a Green Book contract if the lump sum element of the Works is to be provided by a subcontractor under a price-based subcontract (based for example upon the IChemE Yellow or Brown Book) the provisions of Sub-clause 37.3(a) of the Green Book would similarly apply to the cost of Defect rectification, with no further cost to the Contractor and hence to the Purchaser.

If however the lump sum element of the Works were to be undertaken directly by the Contractor, the Contractor's obligation in respect of the cost of Defect rectification would be as set out in Sub-clause 37.3 (b) of the Green Book unless a Special Condition were included in the Contract.

A suitable Special Condition would be:

'Notwithstanding the provisions of Sub-clause 37.3 of the General Conditions, the cost of rectifying any **Defect** in the [*defined Work or element of the Plant to be the subject of the lump sum price*] shall not form part of the **Contract Price** provided the **Defect** is attributable to any breach of the **Contract** committed by the **Contractor**'.

As always the key issue is clarity, while having regard to the hierarchy of the contract documents as set out in Sub-clause 2.2. Hence matters stated in the Schedules should only support the Conditions of Contract, and where anything is to be stated which runs counter in some respect to the General Conditions of Contract those Conditions themselves (through a Special Condition) should be amplified or amended to allow for the proposed entry in a Schedule.

If large elements of the Works are to be paid for on a lump sum price basis, the Guidance on compiling Schedule 19 of the Red Book may be relevant.

Clause 40 provides the facility for a sum being specified in Schedule 18 in respect of any contingency or additional work which may be required. It also provides for prime cost sums to be specified in Schedule 18 in respect of work to be done or Materials supplied by a nominated Subcontractor. Any such sums would need to be clearly identified as such, as would the basis on which payment would be calculated (for example, directly chargeable to the Purchaser at quoted rates or at net cost).

Escalation

Costs which are reimbursed on the basis of actual costs incurred will not require any cost escalation mechanism. However, where Schedule 18 includes unit cost rates it may be appropriate—depending on the length of the contract period—to include a cost escalation mechanism. Clarity as to when such cost escalations would become applicable is essential as well as the method of adjusting the cost rates. Preferably the adjustments would be in accordance with published indices, but if none are appropriate other methods are possible—for example, an auditor's certification of the actual variation of an agreed cost base over the period in question.

Similar considerations apply to any lump sum elements or fees. Normally such elements will be on a fixed price basis but if they are to be subject to adjustment for general price escalation then the applicable method must be stated in Schedule 18.

Guidance on compiling Schedule 19: Terms of payment

See Clauses 39 (Payment) and 40 (Contingencies and prime cost sums) and Guide Note R (Bonds and guarantees)

Clause 39 provides for the necessary certification of and payment to the Contractor of monthly instalments according to forward estimates of his costs, with a reconciliation of the previous month's costs, and all associated processes. These processes are compliant with the Housing Grants, Construction and Regeneration Act 1996. For works falling within the definition of this Act, it should be borne in mind that whenever the full Contract period exceeds 45 days, there is a statutory requirement for interim payments.

If the Contractor is simply to be reimbursed all his costs, Schedule 19 will be brief. For example, it may state: 'The Contractor shall receive his costs and other payments in accordance with Clause 39 and Schedule 18 (Cost elements, rates and charges).'

Such an approach does rely on full and detailed provisions in Schedule 18, including in the case of any fees and other lump sum payments clarity as to when such payments are due. See also the Guidance on compiling Schedule 18 and its references to Sub-clauses 13.2 and 39.2.

Occasionally a Purchaser may require the Contractor to establish a new and separate bank account into which the Purchaser will pay funds for the Contractor then to make payment to his various subcontractors and suppliers. Such a requirement should be set out in a Special Condition.

The nature of cost reimbursable contracts and the payment mechanism set out in Sub-clause 39.3 mean that retention monies under the main Contract are inappropriate. Price-based subcontracts may well have retention monies, but under the main Contract the Purchaser will reimburse the Contractor his net cost incurred at any one time. To do otherwise would oblige the Contractor to fund the cost of the work through the period of the Contract.

Guidance on compiling Schedule 20: Commercial records

See Sub-clauses 3.7, 4.3, 16.10, 19.1, 39.2, 39.6, 42.8 and 43.7

The Contractor is required to compile records from the commencement of the Works, to enable him to present evidence that his expenditure on work under the Contract can be verified to the satisfaction of the Project Manager, and these are required from the time that the Contractor starts work.

The Purchaser may also have to keep records of any costs he incurs which are to be reimbursed by the Contractor in accordance with the Contract. (See Sub-clauses 4.3, 8.2, 10.6, 30.7, 30.8, 33.8, 35.9, 36.4, 37.8, 42.8 and 43.7.)

The parties should have the opportunity to agree that the record-keeping systems they each normally employ should be used for these purposes (or possibly adapted) in preference to creating special forms, etc., at the time that record-keeping is due to start.

Since the Project Manager will be the person most involved in the use of the Contractor's records, he should play a major part in the compilation of this Schedule.

The Schedule should contain, as far as possible, samples of the forms to be used, or in the case of computerised records, printouts of typical records.

In any case, whatever form the records are to take, there will be certain basic requirements. Each document should show:

(a) contract title and identification number;
(b) topic covered, identified in relation to the contract documentation;
(c) date on which the work was carried out or on which the expenditure was incurred;
(d) name of person working, hours worked and resources used by him, as appropriate;
(e) quantities of materials or other resources used, if applicable;
(f) in the case of subcontracted work or purchases from suppliers, the invoiced cost; and
(g) two signatures—the originator's and that of a supervisor of the recording party.

Guidance on certain clauses, the possible need for special conditions, and projects outside the United Kingdom

Guide Note A: Communications

See Sub-clauses 2.6, 11.8 and 12.2

During the performance of the Contract, it is important that all communications should take place in accordance with the requirements laid down in the Contract. A section on Communications should be included in a Project Co-ordination Procedure (see check list in Guidance on compiling Schedule 1 (Description of the Works), item (9) (e)).

Methods of communication can be by secure e-mail, fax, telephone, air or first class mail, courier service or others suited to the parties and their locations. The methods can be extremely flexible, but it is essential that a numbering and recording system is included with notes on persons or companies to which documents are to be copied.

A register for each selected method must be maintained by each party and a hard copy of new register entries should be exchanged on a monthly basis. This is to ensure that a check can be made of receipt by the addressee as well as by other recipients.

Guide Note B: Innovative work or checks by process licensor

Reimbursable contracts are particularly suitable for design of an innovative nature and while in a reimbursable contract the Purchaser normally reviews all the Contractor's work, there may be certain areas where a special review is required by the Purchaser's or external experts.

Examples of these might be the use of catalysts outside their normal operating conditions, or the use of new and previously untried alloys in fabrication. In such circumstances it is advisable to include in the contract a special condition.

In the case of technology licensed from a third party, whether by the Purchaser directly or by the Contractor as part of his Contract, particular care must be taken to define in a Special Condition any role the licensor's personnel may have in checking the Contractor's design documentation and the provision of advice and interpretation of process and engineering information provided as part of the licensor's Documentation or their role in other phases of the project such as training, commissioning or performance testing.

Guide Note C: Intellectual property, know-how, confidentiality and information

See Sub-clauses 4.1 and 43.3 and Clauses 8 (Patent and other protected rights) and 20 (Confidentiality)

The Purchaser may require the Contractor to check information provided by the Purchaser before he uses it, or may wish to rely on some special experience of the Contractor in designing and constructing a particular type of Plant.

In the case where the Contractor's special experience of the technology is involved, a suitable wording for a Special Condition might be:

'The Contractor warrants that he is suitably qualified in the design and construction of plants involving technology of the type to be used in the Plant. The Contractor shall apply his expertise in this respect to the examination of the process design information supplied to him by the Purchaser, and shall immediately bring to the attention of the Project Manager any suspected error, omission or undesirable limitation in such information, and shall assist the Purchaser, as part of the Contractor's responsibilities under the Contract, to improve the process design information to a suitable extent, provided that in so doing the Contractor shall not be required to breach any undertaking given to a third party in respect of proprietary information.'

Any changes to the Specification of the Plant arising as a result of this process should be treated as Variations in accordance with Clauses 16 (Variations) and 17 (Contractor's Variations).

Intellectual Property

Intellectual property is a valuable asset and the owner should identify and protect it by appropriate measures. Various ways of protecting intellectual property are reviewed below.

Patents

A patent is a monopoly granted by the state for a limited period in exchange for the publication of a technical description of the invention. The invention must have novelty, i.e. include an inventive step and be capable of industrial application. Certain types of inventions are excluded by the Patents Act 1977. A patent must be applied for and is granted only after scrutiny to confirm novelty. Fees are required at various stages of the approval procedure. It is granted for five years and can be renewed for a maximum of 20 years.

Copyright

This arises automatically as soon as an original 'work' is written, drawn, recorded, etc and will protect literary work, music, art, film, recordings etc. Software and databases are protected as literary works. Drawings and flowsheets are protected as 'artistic' works (their quality as works of art is not significant). The use of the symbol © is not mandatory under English law but is necessary to establish rights under international conventions in certain jurisdictions. Copyright generally lasts for 70 years and in a database for 15 years from the end of the year in which it was first created.

Registered Designs (Registered Designs Act 1949)
Registration is required. This protects designs which are new and which appeal to the eye. Registered Designs cover surface decoration (sometimes) or objects where the shape is decorative or distinctive (for example domestic electrical goods). Hence this method has little application in process plant construction contracts.

Design Right (Copyright, Designs and Patents Act 1988)
This protects any aspect of the shape or configuration of an original new design. No registration is required and protection is for a maximum of 15 years.

All these above rights enable the owner to take action in the courts to prevent unauthorised use or infringement. The remedies available include injunctions to prevent further infringement, damages to compensate for loss of earnings, including profits made as a result of the infringement and an order to deliver up to the owner of the right any infringing product.

Confidentiality

Ideas protected by the methods discussed above may only constitute a small amount of the organisation's valuable information. The rest of the organisation's knowledge will simply be know-how. In other words, it is treated by the law as simple knowledge. Indeed, all the parties involved in a process project will have information that they wish to be kept confidential.

The Contractor and Subcontractors will have confidential manufacturing knowledge and process know-how. The Purchaser will have confidential information about his own operations and his own processes. All parties will have confidential information of a commercial nature. They all have a considerable interest in protecting the confidentiality of their information.

Sometimes none of the above methods of protection are suitable. This may be because the know-how does not fit under any of them, or because the owner considers that secrecy is a better commercial protection. It is then necessary to require all those who need to know about the product or process to enter into confidentiality agreements. This could include not only employees but also contractors and consultants and their employees.

The confidentiality obligations of both parties are stipulated in Clause 20 which together with Sub-clause 8.6 defines the limitations governing the use by the parties of restricted information and Documentation obtained from the other party. This is meant to cover the parties' own specialist designs and also any know-how which may derive from a process licensor or other outside source.

Sub-clause 8.6 is intended to allow the Purchaser complete freedom to use this information in whatever manner serves the operation of the Plant, including the removal of bottlenecks, maximising output through minor changes and improving efficiency or yields. On the other hand it is not intended to permit the Purchaser to build another plant like, or similar to, the Plant supplied under the Contract and based on the Documentation supplied by the Contractor. In between these extremes, however, it is possible to visualise various major additions that would result in increases in capacity or other improvements. In Sub-clause 8.6 the Purchaser's right to carry out such extensions with the aid of the Contractor's designs is restricted to a capacity increase of 25 per cent. While these arrangements are suitable in a good many instances, there are cases in which the Purchaser's rights in this respect should not be limited at all: for example, if the process design is the Purchaser's own and the Contractor merely provides an engineering service. In other cases it may be appropriate to have a system of further payments related to increased capacity, for example, where a paid-up royalty or its equivalent has been associated with the Contract. In such cases a Special Condition should be written to deal with this and other departures from the provisions of Clause 8 to suit the different circumstances which may apply.

The obligations of the Contractor in relation to information supplied by the Purchaser, as set out in Clause 20, open the way for special measures to ensure confidentiality. This may be particularly important when the Contractor carries out the detailed design or construction of a plant using the Purchaser's own process. In such cases it is sometimes desired to restrict the circulation of Documentation and to ask for return to the Purchaser of all copies of Documentation, etc. at the end of the Contract, or to have other security measures of this kind. Opinions differ on their practical effectiveness but if some deterrent of this kind is required for the Purchaser's protection a suitable Special Condition should be prepared.

The obligations written into the Contract become effective only after the Purchaser and the Contractor have entered into a valid contract. If therefore any confidential information is communicated by either party at the tender stage it should be protected by suitable undertakings given at that stage.

Patents and copyrights

Considerable care must be taken when the Contract is being drafted to ensure that all interested parties that have rights, either under patent or by virtue of copyright, are properly protected. The need for Special Conditions to be drafted in this regard is extremely important, particularly where the Contractor is licensing a process or where a Subcontractor or supplier is responsible for the provision of specialised equipment and know-how. Both primary contracting parties must take every reasonable precaution to ensure that their subcontractors are properly protected under the Contract and that both parties are aware of any implications in the disclosure or use of information being provided one to the other.

The introduction in Clause 8 of the clear vesting of copyright with the Contractor and the Purchaser being granted a licence to use the Documentation has overcome the need to draft Special Conditions clarifying the rights of both parties, though it may be desirable to make some special provision for copyright to pass to the Purchaser in the event of termination for default.

Sub-clause 8.3 limits the liability of the Contractor for any infringement of any patent, registered design, design right, trade mark, copyright or other intellectual property right protected by law to that existing at the date of tender. There always exists a small possibility of a patent being granted to a third party in the period between this date and that of the Plant coming into operation. As the Contract stands, any cost of altering the Plant to avoid infringement or the payment of royalties would be to the Purchaser's account. If the Purchaser is worried about this aspect of the Contract then the use of a Special Condition should be discussed with the Contractor.

If the process is supplied by the Contractor as part of his expertise then it would be reasonable to expect him to review all new patents in the field and inform the Purchaser of any that might be relevant to the design, construction and operation of the Plant and any involving product use. The parties should then discuss what would be the most cost-effective way of overcoming the situation.

A possible division of responsibility that might be agreed as the basis of a Special Condition would be for the Contractor to accept responsibility for the infringement of a design or construction patent while the Purchaser accepted responsibility for patents involving process operation or product use.

Images of the Plant

Sub-clause 20.4 clarifies the question of whether the Contractor has the right to obtain photographs or any other record of the Plant, whether it be a cine film, video recording or some other form of image record. The Purchaser retains the absolute right to allow images of the Plant to be made and to decide whether such permitted images can be used for publicity purposes.

Guide Note D: Unforeseen ground conditions

See Clause 6 (Knowledge of Site conditions) and Guidance on compiling Schedule 19 (Terms of payment))

As with most engineering and construction contracts the General Conditions contain, at Clause 6, an 'unforeseen physical conditions' clause. Strictly the provision will relate to any unforeseen physical condition at the Site, although in practice that is most likely to be an unforeseen ground condition, which may be a natural condition or an artificial obstruction.

Sub-clause 6.2 deems any additional work incurred in dealing with unforeseen physical conditions to be a Variation. The inclusion of such a term is a matter of good management and cost control, as under a cost reimbursable contract the Contractor will be paid his costs anyway.

It is a matter of fact, to be determined by the Project Manager, as to whether particular site conditions are or are not foreseeable. Factors to be taken into account will include the availability of published information on the local ground conditions, what relevant conditions might be inferred from a visual examination of the Site and the extent and reliability of site investigation data made available.

Guide Note E: Taxes

See Sub-clauses 7.4 and 39.10 and Guide Note U (Projects outside the United Kingdom)

The primary intention of Sub-clause 7.4 is to ensure that the Contractor prices and makes allowance for all corporate and personal taxation associated with the performance of the Contract.

With few exceptions all construction work in the UK is subject to Value Added Tax (VAT). Sub-clause 39.10 provides for the normal business approach of VAT being charged in addition to Contract Price.

Guide Note F: Contractor's reports

The Contractor's reports referred to in Sub-clause 3.7 should follow an agreed format including, in addition to the cost control items listed under (a) to (d), up-to-date versions of at least the following:

(a) list of documentation;
(b) approved programme, showing progress;
(c) list of letters, etc. exchanged;
(d) list of new orders, including subcontracts;
(e) list of off-site tests carried out;
(f) list of variations.

Guide Note G: Site services and working conditions

See Clauses 12 (Contract Manager and Contractor's staff), 26 (Health, Safety and Environment), 27 (Site services) and 28 (Site working conditions)

Clauses 12, 26, 27 and 28 deal with the construction Site as regards labour, working conditions and all necessary facilities in a general way suitable for many contracts. However, the details should be included in Schedule 1 (Description of the Works) or Schedule 3 (Responsibilities of Purchaser) bearing in mind the following matters:

(a) Labour and supervision: usually provided by the Contractor;
(b) Housing and transport, canteen or other messing facilities for construction labour: extent to which they are provided and by whom;
(c) Sanitary and first-aid facilities: always required, but may be provided by either party;
(d) Contractor's Equipment, unloading facilities, storage huts or compounds, security arrangements, etc.: usually provided by the Contractor;
(e) Fire-fighting and other safety measures: local circumstances vary widely;
(f) Water, electricity and other utilities for construction purposes: usual for the Purchaser to arrange for supply to a point on the Site and for the Contractor to provide the necessary distribution. Consider the cost and payment (if any), availability (time and capacity) and temporary arrangements;
(g) Consumable materials, utilities or facilities used for the carrying out of take-over procedures: usual for the supplies provided for the Contractor by the Purchaser for construction purposes to continue until the issue of the Final Certificate;
(h) Telephones, fax, secure e-mail and other communications: provision of facilities and payment of cost depend on local circumstances;
(i) Site working conditions: rates of pay and working conditions may be subject to special site agreements between the employers and the trade unions; productivity bonus schemes of various kinds may be stipulated either as part of such site agreements or separately; and
(j) It should be noted that in the United Kingdom and throughout the European Union the construction industry is regulated under the provisions of the EU Directive on Construction. Key aspects of the regulations include:
    (i) safe place of work;
    (ii) precautions against falls;
    (iii) falling objects;
    (iv) work on structures;
    (v) excavations, cofferdams, casions;
    (vi) traffic routes, vehicles, doors and gates;
    (vii) prevention and control of emergencies.

Before construction commences, clear and concise site rules should be drafted and all necessary safety documentation and equipment be made available to ensure safe working conditions at the site. These aspects of safety should be supported by the provision of site induction programmes, safety passport schemes, 'tool box' training and a general ethos of safety. In this regard the Contractor should take a lead in establishing all necessary safety plans but should expect the full support of the Purchaser in the implementation of safety throughout the project.

Many of these details will depend on whether the Site may be considered on its own, is one of a group of adjacent sites where plants are being built more or less simultaneously under different contracts, or is within an established factory.

Guide Note H: Incentives for early completion or improved performance

See Clauses 13 (Times of completion and Approved Programme), 15 (Damages for delays) and 35 (Performance tests)

In cases where it is particularly advantageous to the Purchaser to have the Plant in production or completed as early as possible, it may be appropriate to have a bonus clause. The points of completion to which the bonus is related should be the same as those for liquidated damages and the rate of bonus may be of the same order as the rate for liquidated damages.

In determining the rates of bonus for early completion it should be appreciated that additional production before the scheduled completion date may be less valuable than that after the scheduled completion date. It is probable that the Purchaser has firm orders for his production after the completion date and if no product is available he may have to buy in, possibly at a loss, in order to honour his commitments to long-term customers. On the other hand, production before scheduled completion date may have to be sold at very low margins.

In other cases the Purchaser may find an expanding market is available and consequently wishes to obtain the earliest and greatest possible output over a number of months. This will require discussion between Purchaser and Contractor possibly to adjust the design and to define a bonus scheme related solely to output after acceptance of the Plant by the Purchaser.

As bonuses are the exception rather than the rule they are not included in the General Conditions. If required, a Special Condition should therefore be written; it should state that the Contract Price will include the amount of any bonus and give the method of calculation, the upper limit and any other conditions. The effect of authorised extensions of time should be considered carefully; the bonus may be related either to the original contractual completion date, the point of completion being clearly defined, or to this date plus any authorised extensions. Either basis may be appropriate, but it must be made clear which is chosen. A bonus for timely completion is probably best linked to completion of construction, and not to take-over. The latter is dependent upon a satisfactory outcome to the complexities of the process being brought on stream. Pressure to do things too quickly can have a negative effect on the overall outcome.

If a bonus for timely completion is to be adopted then in addition to stating its value in Schedule 12, a Special Condition should be included in the Contract to provide for such a bonus.

As with proposals for the payment of liquidated damages in the case of failure, any proposal by the Purchaser to pay bonuses should appear in the invitation to tender.

Guide Note I: Variations

See Clauses 16 (Variations) and 17 (Contractor's Variations) and Guidance on compiling Schedule 18 (Cost elements, rates and charges)

While in a reimbursable contract the Contractor is reimbursed for most of the costs he incurs, the Purchaser will have at least a budget cost in mind based on the Contractor's estimate of the work required to fulfil the Specification and complete the work described in Schedule 1 (Description of the Works). It is probable that during the project, change will occur for a variety of reasons; perhaps because the Purchaser's requirements are modified, because the Contractor can recommend improvements from recent engineering experience, or because an alteration is needed in the division of work done by the Contractor under the Contract and other work that has to be done to complete the Plant.

Almost always such change will affect the ultimate cost of the Plant. Therefore meticulous use of Variations to record detail changes of the Specification and the Works is essential to enable both the Project Manager and the Contract Manager to manage the Project and to help avoid disputes.

Clause 16 establishes the need for a Variation to cover change to the Specification or Work and protects the Contractor from major disruptions of his forward business or from having to accept responsibilities which he could not reasonably be expected to bear. The figure of 25 per cent specified in Sub-clause 16.7 may be changed if required by a Special Condition.

Clause 17 caters for Variations proposed by the Contractor. The Project Manager is required to consider such proposals and inform the Contractor whether or not he intends to order such a Variation. The Project Manager's decision is final except where a rejected proposal had the object of eliminating a potential defect or hazard; in that case it may be referred to an Expert under Clause 47 (Reference to an Expert) if there is a dispute.

Any departure from the Specification may have a greater effect than is readily apparent. There have been cases in which severe financial strains have been imposed on either the Purchaser or the Contractor by a series of Variations that have been costed either too high or too low. Experience shows that it is wise to prepare a properly estimated Variation, as a quick approximation usually results in too low an estimate. It is good practice for the Contractor in his accounts to highlight those costs involved in carrying out any Variation.

The limited scope of application of Variations should be noted. Any amendments to the Contract which are not alterations in the Plant, or to the Approved Programme or the type or extent of the Works (for example, changes to any of the Schedules other than Schedule 1 (Description of the Works)) must be made by written agreement. This could simply consist of an exchange of letters.

Guide Note J: Ownership

See Clause 25 (Ownership of Materials)

Clause 25 sets out the primary arrangements as to when ownership or title in the Materials which will form part of the Plant passes to the Purchaser. That the Contract contains such provisions does not automatically mean that title will pass as set out in Clause 25: if under the terms of a subcontract title in certain Materials remains with a supplier or subcontractor then nothing in the main Contract will alter that. Title cannot pass from party A to party B if all the time it is held by party C. It is therefore important that all subcontracts are structured so that title can pass to complement the requirements of Clause 25.

As vesting certificates have on occasions proved to be worthless, the Purchaser would be well advised to ensure that the Project Manager exercises his authority under Sub-clause 25.3. This is particularly so when a stage payment to a subcontractor, made under the terms of a subcontract for a major element of Materials, is to be reimbursed to the Contractor by the Purchaser in respect of Materials which have yet to be delivered to the Site. In such circumstances the provisions of Sub-clause 25.2 are also applicable.

Guide Note K: Insurance

See Clause 30 (Care of the Works) and 31 (Insurance) and Guide Note L (Duties of an employer or occupier)

The Contract requires the Contractor to be responsible for the full care of the Plant and its component items of equipment and other Materials prior to take-over of the Plant by the Purchaser and to a limited extent thereafter. These provisions require the Contractor to make good loss and damage caused while Materials and the Plant are in his care. The Contract includes provisions defining insurance responsibilities and imposing duties to arrange appropriate cover.

Adequate insurance is always necessary, but double insurance (that is, having the same risk covered by two different policies of insurance) inevitably leads to unnecessary expense, and potentially to conflict and counter-claims between the two insurance companies involved. It is generally in the best interests of both Purchaser and Contractor to arrange matters so that one and only one insurer deals with any one area of risk. The General Conditions therefore require the Purchaser to provide Construction All Risks (CAR) insurance for the Plant and Materials until taking over, and also to provide primary cover for the Plant after taking over and for his other property at all times.

The principal objectives behind the form of Clauses 30 and 31 are to ensure that:

(a) Loss or damage to the Plant and Materials before taking over, however caused, will be covered by insurance in the joint names of all parties so that when loss or damage occurs it is unnecessary to apportion fault between the parties involved, and recovery can be made without unnecessary delay to the project.

 The insurance should cover loss or damage occurring during transit. It should also cover damage resulting from defects in design, specification or workmanship, though it would not normally cover the defective items themselves. It should also have a 'reinstatement of sum insured' clause.

 The Contractor must ensure that suppliers and subcontractors provide adequate insurance for equipment and materials while at their premises, unless there is a 'supplier's extension' in the CAR policy. The excess amount, or deductible, payable by the Contractor on any claim must be stated in the Agreement and it will be the Contractor's responsibility to pay this sum on each claim (though he can if he wishes offset liability onto Subcontractors where appropriate).

(b) Loss or damage to the other property of the Purchaser and his Affiliates (including the Plant after taking over) is covered by the Contractor for his fault up to £5 million (£5M) or such other sum as may be stated in the Agreement, and the Purchaser is required to indemnify the Contractor for sums in excess of this amount. The Contractor can, if he so wishes, offset liability onto Subcontractors in any relevant subcontract.

(c) Death or injury to the employees of both the Purchaser and the Contractor (and Subcontractors) for whatever cause is covered by their respective Employer's Liability insurances, so that it is again unnecessary to apportion fault. The intention behind the Contract is that each party and subcontractor involved in the Works should be made responsible for injury to his own employees, even where due to negligence by another party. Clauses similar to 30.6 and 30.7 must therefore be included in each subcontract.

(d) Loss or damage to the Plant occurring after taking over is at the Purchaser's risk, unless it results from something caused prior to taking over or while modifications or repairs are being carried out by or under the supervision of the Contractor, in which case it is covered by the CAR policy. Even if commissioning and performance testing are carried out by the Purchaser without the assistance of the Contractor, it is still advisable for these operations to be covered by the CAR insurance (and therefore in joint names), because, in the event of an accident, it may be alleged that the Contractor was partly to blame as a result of defects in the services prior to take-over. Such cover can also be provided as a standard extension to a CAR policy, whereas it would be unusual in a Purchaser's fire and explosion insurance.

Clause 30 imposes by contract on the Contractor a greater liability for loss and damage to the Plant and Materials than would be imposed by law. A clear insurable interest is created and the financial effects of any voluntarily-assumed enlarged liability are taken care of to a great extent by providing that the Contractor's liability is limited to the replacement cost of the Plant.

Sub-clause 30.7 provides that the Purchaser bears the costs incurred by the Contractor in making good loss or damage to the Plant which is over and above that for which the Contractor is liable under the Contract (defects and deductibles apart).

The provisions of Clauses 30 and 31 are summarised for discussion purposes in Table 3.

Where the parties to the Contract wish to modify these provisions, this may be done by making amendments to the principles set out in Table 3 and passing this information to their professional advisers for the detailed drafting of Special Conditions. The most likely change is that the Purchaser may wish to provide Construction All Risks insurance for the Plant by an extension to his existing Site insurance policy.

The parties must also bear in mind the provisions of Sub-clause 26.5. This provides that, as between the parties at any rate, the Contractor will bear risks arising from environmental pollution and hazardous materials caused by the Contractor. The escape of hazardous or polluting material could easily cause serious losses, both within and beyond the site, and the Contractor would be well advised to check that the insurance cover arranged under Clause 31 includes for these types of loss.

Table 3—Outline of provisions of Clauses 30 and 31

Area of risk	Liability	Limit	Insurance
The Plant, Materials and Documentation before take-over	Contractor to make good all loss and damage prior to take-over (Sub-clause 30.2)	Unlimited	Purchaser to provide transit and Construction All Risks insurance in joint names (Sub-clause 31.1).
The Plant after take-over	Contractor to make good all loss and damage caused by the Contractor (Sub-clause 30.2)	Unlimited	Purchaser to provide transit and Construction All Risks insurance in joint names (Sub-clause 31.1) Purchaser to insure against fire, etc. risks (Sub-clause 31.2)
Other property of Purchaser	Contractor liable for his fault up to £5M**. Purchaser to hold Contractor harmless from sums in excess of £5M** (Sub-clause 30.7)	£5M** (Sub-clause 30.7)	Purchaser to insure against fire, etc. risks (Sub-clause 31.2). Contractor to provide Public Liability insurance of £5M** (with indemnity to principals) (Sub-clause 31.3)
Third party property and persons	Each party liable for its own negligence (Sub-clauses 30.8 and 30.9).	Unlimited	Contractor to provide Public Liability insurance of £5M** (with indemnity to principals) (Sub-clause 31.3).
Contractor's staff and their personal property	Contractor to indemnify Purchaser against all claims (Sub-clause 30.5)	Unlimited (Sub-clause 30.5)	Contractor's Employer's Liability insurance*.
Purchaser's staff and their personal property	Purchaser to indemnify Contractor against all claims (Sub-clause 30.6)	Unlimited (Sub-clause 30.6)	Purchaser's Employer's Liability insurance*.

*No specific contractual provision has been made for these insurances, and they may be covered by a Special Condition, if so desired.
**Or such other amount stated in the Agreement.

Guide Note L: Duties of an employer or occupier

See Clause 30 (Care of the Works)

The following is an outline of some of the responsibilities of an employer.

Liability to third parties

An employer is liable for the acts of his employees in the course of their employment. Cases have clearly established that even if the employee is doing his job in an unauthorised or even a forbidden manner, as long as he is doing what he is employed to do the employer remains liable.

In the majority of cases the employer is not liable for the acts of an independent contractor.

However, in certain cases, he may be liable for such acts. For example:

(a) where he instructs him to carry out the act;
(b) where he has a statutory non-delegable duty;
(c) where he has been negligent by appointing an incompetent contractor;
(d) where the work is carried out on the highway (but not if merely adjacent to the highway);
(e) where the work is particularly hazardous.

Liability to employees

An employer has a general liability in negligence to his workers injured at work as well as statutory duties under the Factory Acts. However, the breach of the general principles laid down in the Health and Safety at Work Act 1974 does not give rise to an action for breach of statutory duty.

Although the state scheme relating to industrial injuries provides for compensation, the majority of employees rely on an action for negligence or for breach of statutory duty since the sums awarded are higher.

The employer has also certain personal duties under law for the protection of his employees which cannot be delegated. This is a complex area and much will depend on the facts of the individual case.

Under normal circumstances the employer will not be liable for injuries at work to the employees of an independent contractor. However, where the employee of the contractor is placed under the direct operating control of the employer, the employer may be liable.

Occupier's liability

The common duty of care owed by occupiers of premises under the Occupier's Liability Act 1954 to lawful visitors applies to employees of both the employer and contractor when on site as well as to third party visitors. Persons other than lawful visitors are covered by the duty owed by occupiers under the Occupier's Liability Act 1984.

Guide Note M: Liability for Defects

See Clause 37 (Liability for Defects)

Clause 37 concerns defects in finished work or in Materials supplied by the Contractor and sets out the responsibilities and obligations of the parties when a defect is discovered. The General Conditions provide that the Contractor is responsible for doing what may be necessary to remedy the Defect at the cost of the Purchaser except where the Defect is made good by Subcontractors at no cost to the Contractor, or where the Defect is due to the failure of the Contractor to exercise due skill and care.

The provisions of Clause 37 regarding the length of the Defects Liability Period and its extension should only be amended if there are very good reasons for having different arrangements.

The guarantee period obtainable from the suppliers of major items of machinery or other equipment may sometimes expire well before the end of twelve months after taking over of the Plant, particularly if such items of equipment have been delivered by the makers to the Site a long time before the whole of the Plant is completed. This possibility must be taken into account by the Contractor during negotiations with the Purchaser on the terms of the Contract.

Certain items of equipment may be designed to have a working life which is short because the capital cost of an item with a long working life would be uneconomic. To avoid disputes about the working life of such items and the suitability of the designs, Schedule 10 (Parts with limited working life) should be included listing the items and showing the estimated working life and also any minimum life guaranteed under normal working conditions.

Any item which fails within the Defects Liability Period, and which has not been specified as designed for a limited working life, may justifiably be considered to have been defective. See also Guidance on compiling Schedule 10 (Parts with limited working life).

There is always a risk of defects arising after the Defect Liability Period has expired and unless steps are taken to exclude them these could give rise to a liability. Under statute and common law an action could be brought for up to six, twelve or even fifteen years after the breach depending upon the nature of the contract. Civil engineering contracts do not generally exclude these matters. However there is a general consensus among process industry purchasers that these statutory and common law rights are inappropriate to the industry.

For the above reasons Sub-clause 38.4 excludes these statutory and common law rights.

Great caution is urged in changing that provision in any way. Nevertheless it is recognised that in matters of civil engineering works such a provision is very different from that industry's normal position, and it may be that in some situations where the Contract includes major civil engineering elements the Purchaser would wish that the Contractor is not relieved of his statutory and common law obligations. If that is the case a Special Condition should be drafted, with particular care being taken to identify the bounds of such a clause. Simple changes to Clause 37 regarding the length of the Defects Liability Period are not an appropriate way to deal with this issue. Equally a general relaxation of the limitation in Sub-clause 38.4, making for an increase in the Contractor's liability, is discouraged for the reasons stated above.

A Purchaser who wishes to benefit from an extended warranty covering one or more items of equipment, despite the general limitation imposed by Sub-clause 38.4, could introduce a Special Condition requiring the Contractor to include in the relevant subcontract either:

(a) a provision that the supplier would be required to provide an extended warranty directly to the Purchaser through a collateral warranty agreement, for which the Purchaser would make an additional payment direct to the supplier; or
(b) a provision that the supplier would be required to provide an extended warranty as a term of the Subcontract in favour of the Purchaser.

The latter approach would draw on the provisions of The Contracts (Rights of Third Parties) Act 1999 and would require clarity of the requirement in Schedule 7 (Subcontracting) (see also Guide Note T (The Contracts (Rights of Third Parties) Act 1999). Arguably the Special Condition should also make clear that the supplier's obligation would apply notwithstanding the provisions of Sub-clause 38.4. While avoiding the need for a separate contract agreement with the Supplier, a Purchaser who adopts this latter approach would be well advised to require sight of the subcontract at an early stage to ensure compliance with the provisions of the Special Condition.

A Contract may be only partly reimbursable and certain of the Contractor's services or Materials supplied may be covered by a lump sum. In that situation it would be appropriate for the Contractor to have responsibilities similar to those in a lump sum contract, i.e. he would be liable for the cost of rectifying Defects in the lump sum element of the Plant. See also Guidance on compiling Schedule 18 (Cost elements, rates and charges) under the sub-heading 'Lump sum items'.

With such an arrangement, the Contractor's maximum liability is likely to be small in relation to the total Contract value and may not fully protect the Purchaser, but it will undoubtedly act as a strong incentive on the Contractor to exercise skill and care in his work. Other arrangements based on this principle are also possible. They may, for example, involve sliding scales, a sharing of rectification costs and of unused contingencies.

Irrespective of the type of incentive scheme adopted based on a 'cost plus fee' arrangement, it is important to distinguish between different classes of defects (to which would attach different financial responsibilities), and to define them properly; this requires careful attention.

For instance, most Contractors would agree to rectify free of charge, i.e. without any addition to the lump sum fee, any defect arising from the work done by the Contractor under this fee but limited to carrying out rectification work of the kind covered by the fee. Thus the cost of 're-engineering' would be the Contractor's responsibility. It is essential to define who should pay for implementing the consequences of such re-engineering.

It is important that the Purchaser and the Contractor agree where the responsibility lies for rectifying different defects. Examples of different defects and their treatment when the scope of the Contractor's input differs are given in Tables 4 and 5 on pages 78 and 79. While these cannot apply to every case it is strongly recommended that they be used as a check-list at pre-Contract discussions so that both parties understand their particular liabilities and the Contractor can fix his price to cover his risk. While it may be attractive to the Purchaser for the Contractor to cover a high degree of risk he should remember the greater the risk covered the higher the Contract Price.

Both Purchasers and Contractors should bear in mind that the above discussion of liability for defects relates only to any liability which may arise between the parties to the contract themselves. In addition to this liability, of course, both Purchaser and Contrac-

Table 4—Examples of liability for Defects or problems

Assumptions

• Project	Process plant, conceptual design and process technology provided by Purchaser.
• Scope of contractor's input	Detail process and engineering design, procurement, construction and pre-commissioning by competent and experienced contractor.
• Contract	Lump sum fee plus reimbursable costs of manpower and material.

Defect or problem	*Rectification*	*Liability for cost of rectification*				
		Purchaser		*Contractor*		*Vendor*
(1) Error in purchaser's information	Unlimited	X				
(2) Unforeseeable ground conditions	More expensive foundations	X				
(3) Drafting errors	Redesign			X		
	Construction rework			X		
(4) Calculation errors	Redesign	X	or	X		
	Construction rework			X		
	Different equipment			X		
(5) Wrong choice of material	Redesign	X	or	X		
	Construction rework	X	or	X		
	Replacement material	X	or	X	or	X
(6) Poor workmanship	Rework			X	or	X
(7) Poor performance of equipment	Replace or modify					X
(8) Wrong specification of equipment	Replace or modify	X	or	X		
(9) Inadequate cleaning during precommissioning	Reclean and repair			X		
(10) Non-specification raw materials		X				
(11) Non-specification services	Plant modifications, additional equipment	X	or	X		
(12) Premature failure of equipment	Repair or replace			X	or	X
(13) Wrong choice of process route	Redesign	X				
	Construction rework	X				
	Different equipment	X				
	Different process	X				

tor may have other potential liability in respect of any defective equipment or Plant which may go into operation. Liability will be primarily of two types—first under appropriate statutes and regulations relating to safety in the workplace (in the UK the Factories Acts and the Health and Safety at Work Act and their resulting codes and regulations), and second under the appropriate laws providing for compensation (damages) for any injury suffered by any employee or other third party as a result.

Some of the defects listed in Tables 4 and 5 cover those affecting plant performance and may only become apparent during the plant performance tests. While the direct cost of rectification is clearly defined, there are additional, often greater, costs of the long periods of waiting by the commissioning team and the construction standby crew, direct labour, waste materials, etc. In the case shown in Table 5, for example, it would be reasonable for the Contractor to be liable for these costs up to the agreed limit, but if so this should be the clear understanding of both parties before signing the Contract.

The Purchaser should note the requirement in Clause 44 (Limitation of liability) that the financial sum for which the Contractor will be held liable must be entered in the Form of Agreement.

Guide Note N: Suspension

See Clause 41 (Suspension of the Works)

There may be any number of reasons for the Project Manager to order the Contractor to suspend all or any part of the Works.

Table 5—Examples of liability for Defects or problems

Assumptions

• Project	Process Plant.
• Scope of contractor's input	Process Licence. Conceptual and detail process and engineering design, procurement, construction and pre-commissioning by competent and experienced contractor.
• Contract	Lump sum fee plus reimbursable costs of manpower and material.

Defect or problem	***Rectification***	***Liability for cost of rectification***				
		Purchaser		***Contractor***		***Vendor***
(1) Error in purchaser's information	Unlimited	X				
(2) Unforeseeable ground conditions	More expensive foundations	X				
(3) Drafting errors	Redesign Construction rework			X X		X X
(4) Calculation errors	Redesign Construction rework Different equipment			X X X		
(5) Wrong choice of material	Redesign Construction rework Replacement material			X X X	 or	 X X
(6) Poor workmanship	Rework			X	or	X
(7) Poor performance of equipment	Replace or modify					X
(8) Wrong specification of equipment	Replace or modify			X		X
(9) Inadequate cleaning during precommissioning	Reclean and repair			X		X
(10) Non-specification raw materials		X				
(11) Non-specification services	Plant modifications, additional equipment	X	or	X		
(12) Premature failure of equipment	Repair or replace			X	or	X
(13) Wrong choice of process route	Redesign Construction rework Different equipment Different process			X X X X		

There may, for example, be business or other reasons or other outside influences beyond the control of either the Project Manager or the Purchaser, which warrant suspension by the Purchaser 'for convenience', or which result from the Contractor's actions or failures in some way. Since, except where he is at fault, the Contractor is entitled to recover additional cost incurred as the result of the suspension, it is clearly not in the Purchaser's interests to have the work suspended except where absolutely necessary; nor is it in his interests to allow suspension to be prolonged, particularly when the cause is within his control. The Project Manager should always consider the impact on the time of completion, cost and any other aspect of the project before issuing a suspension order. Furthermore it must be appreciated that suspension involving interruptions to the supply of some types of major items of equipment may not be practicable.

Both the Project Manager and the Contractor must take notice of procedures and time constraints imposed by Sub-clause 41.3 in order to ensure that neither the Purchaser nor the Contractor is deprived of his rights. It is essential that full and detailed records are maintained throughout the period of suspension in order that additional costs can be substantiated.

Sub-clause 41.4 requires the Project Manager to allow an extension to the project programme by a period of time not less than the duration of the suspension plus any time required by the Contractor to re-establish activities at the end of the period of suspension. If the Purchaser wishes the Contractor to reduce the extension period due to the suspension, it is likely that the Contractor will incur additional costs in accelerating his activities.

Guide Note O: Termination

See Clauses 42 (Termination by the Purchaser for convenience) and 43 (Termination for Contractor's default)

Reimbursable contracts are often used for projects which are not fully defined; for example, the final Specification may be uncertain, supplies of feedstock may not be assured, planning permission uncertain or even the location and capacity of the plant may remain to be fixed. As more information becomes available or economic conditions change, the Purchaser may not wish to proceed with the project and he can, without having to give good reason, terminate the employment of the Contractor under Clause 42.

The General Conditions also provide a right of termination of the employment of the Contractor exercisable by either party in the event of protracted continuance of force majeure circumstances (Clause 14 (Delays)). Furthermore, the Contractor may terminate his employment under the Contract where there is persistent failure on the part of the Purchaser to make due payments (Clause 39 (Payment)) and where there is prolonged suspension of the Contractor's performance of his obligations by order of the Project Manager (Clause 41 (Suspension of the Works)).

Clause 43 enables the Purchaser to terminate the employment of the Contractor in the event of the Contractor going into liquidation or the like, or is in default as described in Sub-clause 43.2.

In all cases of termination, the position of the parties is regulated by the provisions of Clause 42 or 43.

Sub-clause 42.10 or 43.11 as applicable identifies those provisions of the Contract which survive the termination.

Guide Note P: Limitation of Contractor's liability

See Clause 44 (Limitation of liability) and Guide Note M (Liability for Defects)

Although rare in traditional building or civil engineering contracts, waivers or exclusions of liability for economic, indirect and other 'consequential' loss are invariably a feature of contracts for work in the process plant sector. The primary reason is one of scale: the Purchaser's capital investment and production costs (and frequently profits) in the industry (particularly in the oil, gas and pharmaceutical sectors) are so high that it is neither reasonable nor practicable for the risk of such losses to be imposed upon the Contractor who is incapable of carrying it and, what is more, cannot normally insure himself against it.

Thus there is a general consensus among Purchasers and Contractors alike that it is impractical to expect the Contractor to carry the risk of consequential loss associated with the design and construction process in the industry. This risk should therefore be taken into account by the Purchaser in making his original investment decision.

There are a number of specific limitations of liability set forth in Clause 44 and there are provisions in the Agreement for maximum monetary limits to be placed on the Contractor's liability to the Purchaser for defaults under the Contract. However, users of the Conditions should note the last sentence of Sub-clause 44.2. The purpose of this is to make it clear that in incorporating some limitations of the Contractor's liability there is no intention to cause the wording of the Clause to be in conflict with the requirements of current 'consumer protection' legislation within the United Kingdom. The provision makes it clear that the limit of liability stated in the Agreement does not in any way affect any liability of the Contractor to third parties under general law.

The Unfair Contract Terms Act 1977 renders void any contract term which purports to exclude liability for death or personal injury. It also renders void any notice which purports to exclude liability for death or personal injury due to negligence. In non-consumer contracts a term excluding liability for damage to property is subject to a test of reasonableness and Schedule 2 of the Act provides guidance on what is to be considered reasonable.

The Consumer Protection Act 1987 imposes strict liability on manufacturers for any injury to a consumer or his property caused by a product. Unlike a claim in tort there is no requirement for proof of negligence. At first sight this would not appear to be relevant in a contract for the construction of plant but the Act permits the injured party to claim against a wide range of parties in the supply chain or even against producers of a component of the product. If the Contractor operates the plant for a period before handover and produces product which is then sold to consumers by the Purchaser, the provisions of the Act could apply to the Contractor. In effect, therefore, this Act creates a new basis for claims by any person injured by defective equipment directly against the manufacturer, and the Contractor and Purchaser should pay particular attention to ensuring that adequate instructions for use, operation and maintenance are issued with equipment and that adequate inspection, quality control and records are maintained at all times.

Contractors should also note that all other countries within the European Union now have legislation in place which is for all practical purposes identical to the Consumer Protection Act. At least a dozen other countries have also enacted similar legislation.

It should be noted that while Sub-clause 38.4 provides that the Contractor has no further obligations to the Purchaser following the issue of the last Final Certificate, such a contractual provision would not protect the Contractor against any third party claim under, for example, the Consumer Protection Act 1987.

Guide Note Q: Dispute resolution

See Clauses 45 (Disputes), 46 (Adjudication), 47 (Reference to an Expert) and 48 (Arbitration) and Sub-clauses 14.1, 16.7, 16.9, 17.3, 17.4, 18.3, 21.4, 32.7, 33.7, 33.10, 35.8, 37.9, 38.5, 41.5, 42.5 and 43.9

Care should be taken in a reimbursable contract not to create a dispute over a matter which is simply a normal error or omission. It is accepted that errors and omissions will occur and that they are to be rectified at the Purchaser's expense, unless they arise from the failure of the Contractor to exercise proper skill and care.

The Contract places a duty on the parties to use all reasonable endeavours to avoid disputes (Sub-clause 45.1) and, in the event a dispute does arise, to attempt in good faith to reach a settlement by negotiation (Sub-clause 45.6). This underlies a key part of IChemE's contract philosophy that a project is best served by teamwork and co-operation between the parties, not by confrontation.

In the event that a disagreement cannot be resolved by negotiation, steps are set out (Sub-clauses 45.3 and 45.4) which must be followed by the parties if a dispute under the Contract is to be deemed to have arisen. Short-cutting or ignoring these requirements and a further series of requirements set out in the appropriate IChemE or other Rules (see below) may lead to legal problems later when a particular form of dispute resolution has been chosen.

The Contract provides (Sub-clause 45.7) for a dispute to be referred by agreement for mediation, for example to CEDR or the CIArb if that is what the parties wish.

The General Conditions provide a variety of forms of dispute resolution in its Contracts to meet the needs of different types of dispute. IChemE also publishes Rules for the conduct of Expert Determination, of Arbitration and of Adjudication which are incorporated into the Contract as the default position.

But by far the best approach is to avoid disputes.

Expert Determination

A number of matters such as quantification of disputed Variation Orders are identified in the Contract as being matters which may be referred to an Expert for determination. Matters so identified may be referred by either party without the agreement of the other party. A referring party in these circumstances is not obliged to adopt this method of dispute resolution and may instead invoke any other method for which the Contract provides, including arbitration and, in appropriate circumstances, adjudication.

Provision is also made for any other matter in dispute to be referred to an Expert for a final decision but only by agreement between the parties.

Once a dispute has been referred to an Expert, it ceases to be referable to arbitration (Sub-clause 47.7). Parties should, therefore, be careful if there are a number of issues likely to need resolution not to arrive at a situation where some go to Expert determination and some to arbitration, which would be undesirable because inconsistencies may result.

The powers of an Expert are set out in the IChemE's 'Rules for Expert Determination'. These also set out in detail the procedure to be followed by the parties in seeking the appointment of an Expert.

An Expert's findings are final, conclusive and binding on the parties (Sub-clause 47.7). Subject only to his either having failed to answer the questions put to him or having failed materially to act impartially between the parties, the Expert's findings cannot be appealed to the Courts.

Expert Determination is particularly suited to those disputes where the knowledge and experience of a senior and respected practitioner in the field is likely to assist in reaching a speedy and fair result.

Arbitration

In contrast arbitration, although it has some similarities to Expert Determination, is carried out under the provisions of the Arbitration Act 1996*, as detailed in the IChemE's Arbitration Rules. The 1996 Act has moved a long way towards making English Arbitration awards final, by reducing substantially the number of matters that may be taken to the courts on appeal. Apart from 'partiality' by the arbitrator or his failing to answer the right questions, essentially the courts will only interfere on points of law and then only on points seen as substantially affecting the rights of the parties or as being of general public importance. Although Sub-clause 48.4 appears to exclude all appeals to the Court, in practice it would probably only exclude an appeal on a point of law under s69 of the Arbitration Act 1996, and challenges on points of jurisdiction or serious irregularity under SS67 or 68 would probably be heard.

The 1996 Act, in parallel with modern court procedure, places a duty on the arbitrator to 'manage' the reference efficiently and gives him substantial powers to enforce his duty, partly through the costs mechanism and partly through the power to disallow evidence submitted late.

To cater for different types and magnitudes of disputes, IChemE has provided in its Rules for a variety of forms of arbitration (documents only, procedure for expert witnesses, etc.) as well as a full, formal procedure. The choice of the most appropriate form will be discussed by the arbitrator with the parties at the first meeting for directions.

If the parties do not wish to use IChemE's Arbitration Rules, they may by agreement use some other set of rules, such as CIMAR (Construction Industry Model Arbitration Rules). Any decision to change the 'default' rules in the Contract should be taken at the time of contract negotiation. Once a dispute has arisen such changes are very difficult to agree. The parties should also be aware that IChemE's Rules are consistent with IChemE Contracts and the Arbitration Act 1996. Choice of another set of Rules may lead to inconsistencies later.

Clause 48 constitutes a proper Arbitration Agreement as required by the 1996 Act. Changes to the wording of Sub-clauses 48.1 and 48.2 in particular should only be undertaken with legal advice.

Adjudication

If a review of the Works and the Plant leads to the conclusion that the Contractor will be required under the Contract to undertake activities which are defined as construction operations by the Housing Grants, Construction and Regeneration Act 1996 Part II, Clause 46 will apply to disputes arising from those activities, as will the IChemE Adjudication Rules. Whilst it is hoped that parties will generally follow Sub-clause 45.4, Sub-clause 46.2 makes it clear that for activities which do fall under the Act, either party may call for adjudication at any time without first formally establishing a 'Dispute'.

Clause 46 has been modelled to correspond with Clause 108 of the Housing Grants, Construction and Regeneration Act 1996 under which an adjudicator's decision is binding until the dispute is finally determined by legal proceedings, by arbitration or by agreement. Sub-clause 46.8 of the Contract mirrors section 108(3) of the Act.

While there is, as yet, no judicial authority on the point, it is arguable that adjudication under the Act can proceed even if an Expert Determination is also in hand. If so, it would appear that an Expert's finding given before an adjudicator has given his decision will stay the adjudication, while a finding given after a decision will over-ride it. But it would be far better to avoid the problem by choosing one approach or the other at the earliest possible stage.

Guide Note R: Bonds and guarantees

Sometimes the Purchaser requires a guarantee or guarantees from a third party that the Contractor will perform the Contract and/or repay any money paid in advance if the Contract is terminated or otherwise not performed. The third party may be one (or more) private persons or a bank, insurance company or business specialising in the giving of such guarantees. These are generically referred to as bonds, but may take the form of insurance policies, bank guarantees, or repayment performance

* in England, Wales or Northern Ireland. Similar provisions (whether by statute or in common law) apply elsewhere.

bonds. There is some confusion of thought about these documents because different forms are required for different purposes; the basic purpose however is always to provide security to the recipient. Giving the bond or calling for payment under it does not affect the rights and obligations of the Purchaser or Contractor under the Contract or in respect of its breach or termination.

In general the procedures and forms used by the banks are followed to a greater or lesser extent by the other bodies which provide such guarantees. In the following part of this Guide Note the words 'bank' and 'bond' refer to all givers of guarantees and the guarantees given respectively.

The provider usually insists upon:

(a) The maximum amount payable being expressed in both words and figures.
(b) A calendar date for expiry of the guarantee (a set time, such as three hundred and sixty five days after taking over, will not be acceptable because the bank cannot know for sure the starting date of the three hundred and sixty five days). To allow for time to call the bond, the expiry date should be thirty days after the latest date the obligation guaranteed is likely to expire. The cost to the Contractor will of course depend on the length of time the bond is in effect.
(c) A statement that the guarantee is irrevocable and unassignable.
(d) An obligation to pay on receipt of the appropriate document(s) without any requirement or right on the part of the bank to enquire into the reason, reasonableness or justification for calling the bond.
(e) A requirement to return the bond for cancellation when it has expired.
(f) A law and exclusive jurisdiction clause.

The bank will not enquire into or adjudicate upon whether the Contract has been correctly performed. The bank's duty is merely to see that the demand is in proper form according to the tenor of the bond, and is regular on the face of it. On receipt of such a document the bank is bound to pay, unless it has notice of provable fraud, and will certainly have an indemnity whereby it can recover the money from the Contractor. (Where, however, the 'bank' which gives the bond is not a bank but some other financial organisation, payment may not be made upon first demand.)

It follows that the wording of the operative part of the bond must describe exactly the document(s) to be presented as, or with, the demand for payment so as to reflect precisely the purpose of the bond and the intentions of the parties. If the bond is to be given during the effective period of the Contract, a Special Condition will be needed stating when the bond is to be given, how the wording is to be settled, and the terms listed in (a) to (e) above.

An appropriate Special Condition might read:

'On or before .... the Contractor shall at his own expense furnish to the Purchaser an irrevocable guarantee/bond from a surety who is a first class bank or insurance company or other surety or sureties acceptable to the Purchaser (such acceptance shall not be unreasonably withheld) in an amount of £......... [and in words]. Such guarantee shall:

(a) Expire not later than ....;
(b) Provide payment on written demand of the Purchaser stating ....;
(c) Be binding notwithstanding such Variations, alterations or extensions of time as may be made, given, conceded or agreed under these Conditions of Contract;
(d) Be returned to the surety on expiry.'

It may be desirable for the Special Condition to state what the consequence will be if the bond is not furnished.

The purposes for which bonds are required vary considerably but the following are common.

Tender bonds

Tender bonds, also referred to as bid bonds, are not within the scope of the Contract, as they are required at the tendering stage. The purpose is to ensure that the successful tenderer enters into a Contract or pays the Purchaser his expenses of abortive negotiations. It may be payable on a written demand stating that the tenderer was informed that his tender was acceptable but nevertheless within a stated number of days failed, neglected or refused to enter into a binding Contract.

Down payment or advance payment bonds

The purpose is to secure repayment of the down or advance payment (or a suitable part thereof) if the Contract is terminated, frustrated or otherwise comes to a premature end. Payment may be on written demand of the Purchaser without proof or conditions but the amount payable should reduce in steps by stated amounts or percentages as the Contract proceeds, normally in proportion to instalments of the Contract Price payable and at the dates the instalments are due. A Special Condition should specifically impose upon the Purchaser the obligation to inform the bank (or other surety) whenever a Contract instalment is due and the down payment bond has been reduced in face value. This obligation should take place within three to five working days and should be communicated by copy to the Contractor.

These bonds are not normally appropriate to cost reimbursable contracts except where the Purchaser is reimbursing the Contractor a payment made for Subcontract Materials which remain off-site (perhaps because they are only partially manufactured).

Performance bonds

The purpose is to secure due and faithful performance of the Contract by the Contractor and to provide the Purchaser with funds to set about getting the Contract performed elsewhere and/or to meet damages if the Contractor fails to perform. The amount is usually five to fifteen per cent of the original Contract Price. Payment is on written demand including in full detail of all particulars known to the Purchaser of the Contractor's failure to perform and stating that the Contractor has been given so many days notice of the Purchaser's intention to call the bond. The Purchaser and the Contractor should agree by a Special Condition that the validity of the performance bond will be extended beyond its original expiry date at the sole option of the Purchaser and at the cost of the Contractor if the Contract has not been performed by the expiry date.

This may also be used to cover the payment of any liquidated damages due under the Contract.

Retention bonds

See Guidance on compiling Schedule 19 (Terms of payment)

Purchasers would usually prefer to make the final payment after the results of performance tests are known and after the Contractor is finally released from his liabilities for Defects. It is recognised that this adversely affects the Contractor's cash flow and unreasonably prevents him from realising his profit when all, or virtually all, of his work has been done. It is therefore common practice to pay the final instalment on taking over or completion of commissioning provided that the Contractor furnishes an 'on demand' bond for an amount equal to the final payment with an expiry date no earlier than the date on which the Contractor is released from his liabilities for Defects.

Like retention monies, these bonds are not appropriate to cost-reimbursable contracts (see Guidance on compiling Schedule 19 (Terms of Payment)).

Parent company guarantees

Parent company guarantees may be required when a party to the Contract is a subsidiary or a company specially set up for the Contract (for example, a jointly owned consortium company or a company to trade in a particular country). Assistance in obtaining credit ratings can be provided by commercial banks.

Guide Note S: Completion, taking over, testing and start-up

The concept of completion by the Contractor of the work he has undertaken to carry out under the Contract has several aspects of significance:

(a) compliance with the Specification and with the description of the Works;
(b) finishing particular tasks in accordance with the Approved Programme, or with reference to financial incentives;
(c) transfer of responsibility.

The General Conditions have been devised to cater for each of these by both procedure and documentation. A summary is given in Table 6.

The conditions do not use the term 'Mechanical Completion' because process contracts include work of many different engineering disciplines and to concentrate upon any single discipline would be misleading.

For rather different reasons, the civil engineering concept of 'Substantial Completion' is not used.

Mechanical completion is therefore included within Completion of Construction, which should be recognised as when everything described in the Specification is physically complete, and also when any prescribed tests have proved successful (see Guidance on compiling Schedule 14 (Criteria for the completion of construction)).

The terms 'start-up', 'pre-commissioning' and 'commissioning' have not been used in the Conditions. They are not sufficiently well defined to enable the Project Manager to issue a certificate with a precise date (see Guidance on compiling Schedule 19 (Terms of payment)).

Completion of Construction is one important break point. After it has occurred the Contractor's work of take-over testing should not be rushed, because it could affect the safety of the Plant. For that reason, Completion of Construction should preferably be used in Schedule 11 (Times of completion) as a point at which any delay results in payment of liquidated damages, if they are to be applied. The criteria which are to be satisfied by the Contractor before the issue of a certificate are defined in

Table 6—Procedures and documentation for completion

Topic	Clause of General Conditions	Document defining contractual requirement	Record of completion	Record document prepared by*
The Plant	32.2 32.3 32.4	Specification: Schedule 14 (Criteria for the completion of construction)	Construction Completion Certificate	Contractor (draft); Project Manager may endorse. Project Manager to finalise
Time	32.6	Schedule 11 (Times of completion)	Construction Completion Certificate	Project Manager
Responsibility for the Plant	33.7	Schedule 15 (Take-over procedures)	Take-Over Certificate	Project Manager
Plant performance	35 36	Schedule 16 (Performance tests and procedures) Schedule 17 (Performance guarantees and damages for failure)	Acceptance Certificate	Project Manager
Final completion and end of final Defects Liability Period	38	Specification; Schedule 1 (Description of the Works)	Final Certificate	Project Manager

*All certificates are issued by the Project Manager to both Purchaser and Contractor. A sample of each certificate is shown in the Appendices at the end of these Guide Notes as follows: A. Construction Completion Certificate; B. Take-Over Certificate; C. Acceptance Certificate; D. Final Certificate.

Schedule 14. However, if the Purchaser wishes to have liquidated damages applied to delay in taking over, the procedure to be followed is described at the end of Guidance on compiling Schedule 11.

It is very important from both contractual and safety points of view that responsibility for the Plant should pass from the Contractor to the Purchaser in a clearly defined way (see Sub-clause 33.7). The time for this to be done is preferably just before any raw material is first fed into the Plant to undergo the designed process changes. However in some circumstances the Contractor may take responsibility for the initial operation of the plant during a take-over test.

It should also be noted that the Defects Liability Period starts on the date of taking over. From that time the Purchaser is responsible for the care, operation, maintenance and safety of the Plant.

Except in the most simple of projects, the Plant will consist of several sections, some of which are operationally dependent on others. There will thus be a logical sequence in which completion of each will be required in order to achieve the most satisfactory commissioning of the whole project. Furthermore, the actual task of thoroughly inspecting and testing a completed plant before it can be used safely can be costly in time and resources. To leave the whole of a project to be inspected and tested at the same time could lead to the procedure being rushed because of pressure for production to start (and hence to potential hazards), or else to delay if the inspection work is to be done properly. For these reasons the Approved Programme should specify the order of completing the different sections in some detail, and should allow adequate time for inspection and testing.

Sub-clauses 32.1, 33.1, 35.1 and 36.2 allow for sections of the Plant to be specified for completion of construction, taking over and acceptance respectively.

When agreeing on such sections, which should be shown in the Specification and then referred to in Schedules 11 and 12 (Liquidated damages for delay) if liquidated damages for delay are to apply, the parties should consider the following:

(i) Some plants will lend themselves to being divided into sections. For example, the process unit in which the final product is formed will probably depend on other units for inputs including feed stocks, intermediate products and utilities, each of which could be taken over as a section. The final product is likely to flow to a storage unit and outloading facilities or packaging, and effluents may require separate treatment units, which could also form defined sections.

(ii) The advantage of specifying such sections lies in the opportunity to concentrate the efforts of the personnel involved in attaining and checking completion criteria, take-over procedures or performance tests on one or two sections at a time.

For each section of the Plant the following list, although not complete or appropriate for all projects in every respect, may be used as *an aide-mémoire* in identifying those activities which should be included in Schedule 14 as part of construction (possibly up to and including item (k)), and those activities which should be classed as take-over procedures (possibly (l) to (s)). The remainder could then be carried out by the Purchaser and his staff after taking over with advice from the Contractor, who should be able to observe the actions of the Purchaser sufficiently closely to satisfy himself that his operating and maintenance instructions, where applicable, are being observed:

(a) Handing over of Test Certificates (in particular those required by statute) relating to any item of Materials.
(b) Installation of gland packing and lubrication of valves and minor machinery; checking rotation of drivers.
(c) Isolating equipment, removing expansion joints, rupture discs and relief valves from pipework and equipment before line flushing, cleaning and testing and reinstalling thereafter.
(d) Cleaning pipework by flushing, steaming, blowing, pigging or scatter scaling.
(e) Hydraulic and/or pneumatic pressure testing of equipment and pipework.
(f) Applying special treatment or other preparation of inside surfaces.
(g) Testing relief valves.
(h) Removing loose material and dirt, installation of internals, inspection and closing up of vessels.
(i) Installing orifice plates after line cleaning, installation of temporary strainers, removal of slip plates.
(j) Checking electrical installations.
(k) Checking alignment and direction of rotation of major machinery and drivers; cleaning and flushing, cooling and lubricating oil installations.
(l) Conducting short running-in tests on machinery.
(m) Checking action of instruments and control valves, testing electrical controls, alarms, computer systems and software.
(n) Calibrating instruments.
(o) Drying out or purging plant.
(p) Drying out refractories.
(q) Adjusting pipe supports for expansion.
(r) Removing, cleaning and replacing temporary strainers.
(s) Checking orifice plates and permanent blanks, relief systems, drainage systems, safety provisions.
(t) Charging catalysts.
(u) Charging raw materials, process chemicals, fuel.
(v) Warming up, starting up fluid flows.
(w) Tightening hot joints.
(x) Starting up and operating the various sections of the Plant.
(y) Routine maintenance, cleaning, plant adjustments.

Some of these actions may reveal construction errors, which will then be rectified while the other activities mentioned are proceeding. Insulation and painting work is also likely to go on at this time, together with minor construction details not previously finished.

Plant start-up and performance tests

For the majority of process plant projects the later phases of construction, the preparation for start-up and the early phases of operation proceed more or less continuously without clear boundaries. The work done in this transition period is customarily called 'commissioning', but as this is an ill-defined term it is not used in the General Conditions. Nevertheless commissioning is an essential part of the work required to provide a completed process plant which operates in accordance with its designer's intentions. It is usual for the first attempt to start-up a plant to reveal weaknesses or omissions, however small, and it is therefore customary to plan the start-up in stages, so that if a shut-down or pause is required for rectification or adjustment, this can be done in an orderly manner at the minimum cost of time and resources. To be successful a plant must not only be mechanically complete and tested but it must also have feed-stocks, intermediate chemicals and utilities and must be served by properly trained operating and maintenance teams. The organisation of the commissioning period within an agreed procedure is very important; responsibilities, particularly those relating to the person in charge of commissioning, must be clearly defined and lines of communication must be set up.

IChemE publish a guide entitled 'Process Plant Commissioning' (available from IChemE Book Sales).

The advantage of the Purchaser's staff having the responsibility of operating and maintaining the Plant from start-up is that it forms a culmination of their training and makes them concentrate on following meticulously the Contractor's instructions and the procedures specified in the operation and maintenance manuals, which will incorporate information provided by Subcontractors and suppliers. If they fail to follow these instructions and the Contractor can show such failure, then he may make out a case to be excused from achieving the guaranteed results in a particular test or he may be able to show that a Defect which appears is not his responsibility.

Sub-clause 35.4 provides that the performance tests are to be carried out by the Purchaser in the presence of the Contractor, with the operating and maintenance staff needed to start-up and operate the Plant being provided by the Purchaser. This has generally been found to be the best arrangement when the process technology and performance guarantees are provided by the Contractor. As a rule the personnel required will have been selected and trained during the period of procurement and construction of the Plant and arrangements will have been made for specialised training by the Contractor or by equipment suppliers or licensors (in which case such arrangements should be described in Schedule 9 (Training by Contractor)). The Contractor's staff who are present during the Plant start-up support the Purchaser's staff and assist in decision-making.

In any case, it is in the interests of both parties for the Contractor to be present until the performance tests have been successfully completed so that the Plant is safeguarded against damage by faulty operation, which, if caused by the negligence of the Purchaser's staff, may invalidate any guarantees given by the Contractor and supplier's warranties.

It is not unusual for the Contractor to provide for such experienced staff to remain on site at the request of the Purchaser, for as long as he may require so that they can be released when the Purchaser's own staff are confident enough to proceed without them.

Guide Note T: The Contracts (Rights of Third Parties) Act 1999

This Act, which applies to England, Wales and Northern Ireland, allows a third party to a contract (i.e., someone who is not directly a party to it) to exercise any rights which the contract might confer or purport to confer upon him.

In the context of a contract between a Purchaser and a Contractor, any Subcontractor is a third party and thus may be able to claim that the contract gives him certain rights which he may wish to exercise directly. In the same way, in relation to a Subcontract, the Purchaser would be a third party. This is a change to the Common Law doctrine of privity. Note that the Act does not in any way confer obligations on a third party—obligations can only be imposed upon someone who is directly a party to a contract.

The Act also allows for the inclusion of words into a contract so as to exclude the effect of the Act.

The latter approach has been partially adopted in this edition of Green Book. Sub-clause 9.12 renders the Act inapplicable except to the extent that any provision in Clause 9 (Assignment and subcontracting) permits otherwise.

Sub-clause 9.11 expressly permits any Subcontractor to exercise rights directly against the Purchaser in respect of any misuse of Confidential Information and Sub-clauses 9.7 and 9.10 permit the Purchaser to enforce the terms of any subcontract guarantees.

These rights are in addition to any rights of action between the Purchaser and Contractor.

The advantage is that the parties most directly affected (which may be the Purchaser and the Subcontractor) can deal directly, for example where there has been a breach of a confidentiality undertaking, or where an item of plant supplied by a Subcontractor develops a Defect.

Guide Note U: Projects outside the United Kingdom

As written, the Form of Contract is subject to English law, and is concerned with a project carried out within the United Kingdom. However it has become increasingly popular over recent years for use in connection with projects outside the United Kingdom.

This raises two separate, but related, issues. First the parties may wish to change the 'governing law' of the contract from English law to the law of some other country. Second the contract may well need to take account of the fact that many legal requirements relating to a Plant site outside the United Kingdom will be very different from those which apply to a Plant site within the United Kingdom.

Any Contractor or Purchaser planning to undertake a project outside the United Kingdom should take practical advice from organisations which are experienced in the legal, commercial, taxation and technical requirements and 'custom and practice' in the country or countries in which the Works are to be executed.

The sufficiency of any lump sum prices for subcontracts is particularly important when the Plant is to be constructed in a country foreign to the Contractor and/or Purchaser. The Contractor, when working outside his own country, is always considered to be the party at greatest risk. However, if it is decided to employ this form of contract the sufficiency of a local contractor's tender price must be fully investigated before any contract is signed. It must always be recognised that the practices which apply to commercial arrangements vary considerably from country to country (see also Guide Note E (Taxes)).

Clause 7 (Statutory and other obligations) is sufficiently general to be applied to any country in which the Contract is to operate. The importance of Sub-clause 7.4 is highlighted because in an international project which may be carried out in more than one country, the complexity of the various tax systems, Value Added Tax (VAT) charges and currency exchange regulations may have a significant effect on the commercial success of the project for either or both parties. Where specific laws, regulations, rates, taxes and any other factors apply which are likely to have an effect in time or cost, these should be clearly stated in a Special Condition.

The subject of taxation for projects outside the United Kingdom is often very sensitive and will always require the most thorough investigation by Purchaser and Contractor alike. It is clearly beneficial for the Purchaser that the effect of taxation upon the cost of the project is minimised and, therefore, agreement in respect of responsibility must be of the highest priority. It may well be possible greatly to reduce the tax burden by the use of an 'offshore' and an 'onshore' contract. These will cover work carried out remote from the country in which the project is located and that carried out inside the country respectively.

At the time of publication all new Plant work carried out in the United Kingdom and goods for export are zero rated for VAT purposes. For projects within the European Union it is important for the Contractor to investigate thoroughly the VAT rates which apply, both to any equipment and services purchased by the Contractor outside the United Kingdom, and to the Contractor's price. These may well vary, depending upon which countries are used as sources for the project. For projects outside the European Union it is important that the Contractor should check what taxes might apply.

Import taxes are levied by various countries and they may vary in percentage terms depending on the category of equipment being supplied. They are normally payable in local currency at the time of importation through customs.

Withholding tax is levied by some countries. This is based normally on payments made to the Contractor by the Purchaser as a tax on the profit, or 'deemed profit' made by the 'foreign' supplier. Payment may be required in convertible currency by the authorities, or alternatively money may be deducted from payments due. The level at which withholding tax is levied may be very high (as much as 30 per cent of the Contractor's 'deemed profit') and it is therefore essential that very great care is taken to establish the extent of such liability as well as the extent (and the time required) to which recovery of withholding tax may be made either directly or under any double taxation agreement.

Income tax for the Contractor's staff working in overseas countries is normally the responsibility of the individual employee. However it is recommended that the Contractor undertakes a review of this tax as disruption of the programme can result when employees require home leave to avoid liability for foreign taxes.

Clause 28 (Site working conditions) recognises that the Contractor as well as the Purchaser must obey the laws, regulations and customs and practices that apply to the Site. Attention is drawn to the check-list given in Guide Note G (Site services and working conditions) which includes some of the important factors which need to be taken into account. Nevertheless, the Contractor must satisfy himself that he is fully aware of all the conditions that exist which could affect his price or his ability to complete the Works in accordance with Schedule 11 (Times of completion).

Insurance needs particular attention when a project is to be carried out in countries outside the United Kingdom and local advice should be taken. Professional indemnity or product indemnity insurance policies often have very strict limitations on the countries for which cover is provided. Contractors would be well advised to investigate the coverage and premiums for all applicable insurance policies when working outside their own country. (See Clause 31 (Insurance) and Guide Note K (Insurance)).

Contractors would also be well advised to investigate local health and safety practices and the possible consequences both corporate and individual of accidents, particularly fatal accidents.

Sub-clause 44.2 refers to the limitation of the Contractor's liability in the event of a breach of contract. This Sub-clause then goes on to refer to the Consumer Protection Act 1987 and the Unfair Contract Terms Act 1977 which are only relevant to contracts operating in the United Kingdom. It must be recognised by the parties to the Contract that there may be particular laws and statutes which have the effect of overriding the limitations of liability set in the Contract (see also Guide Note P (Limitation of Contractor's liability)).

See page 87 for a draft agreement in a form suitable for international application. This contains several items referring to Sub-clauses in the Conditions, and since the Agreement takes precedence over the Conditions (see Sub-clause 2.2) these would have the same effect as individual Special Conditions.

The following Clauses of the General Conditions should be given particular attention, because most will need to be deleted, replaced or modified by the inclusion of suitably worded Special Conditions.

Sub-clause 2.1: law.
Sub-clause 7.4: taxation (See Guide Note E (Taxes)).
Sub-clause 9.12: rights of third parties.
Sub-clause 26.1: Health and Safety.
Sub-clause 27.5: certificates required by law.
Sub-clause 31.3: third party insurance cover in pounds sterling.
Sub-clause 39.10: Value Added Tax.
Clause 46: adjudication under English law.

In addition all references in the General Conditions to Adjudicators or Adjudication should be deleted (see Sub-clauses 12.5, 17.3, 28.4, 42.10, 43.11, 45.2, 45.8, 46, 48.1, 48.3.)

Draft Agreement suitable for projects outside the United Kingdom

The draft agreement that follows is of a very simple nature. It may well need considerable modification or expansion (together with the addition of Special Conditions) to fit the needs of a particular project.

Agreement

Contract Title and Identification Number:

THIS AGREEMENT is made the *[day, month and year in words]*

between

of

(hereinafter called 'the **Purchaser**') of the one part

and

of

(hereinafter called 'the **Contractor**') of the other part.

WHEREAS

The **Purchaser** wishes to have a process plant to be known as

constructed at *[location including the name of the country]*

and wishes the **Contractor** to carry out and complete the **Works** as defined in the **Contract** and the **Contractor** is willing and able to carry out and complete the **Works** in accordance with the **Contract**.

THIS AGREEMENT provides as follows:

1. The following documents and their attachments (if any) shall together constitute the contract between the **Purchaser** and the **Contractor** and the term '**Contract**' shall in all such documents be construed accordingly.

   (a) This Agreement.
   (b) The Special Conditions (if any).
   (c) The General Conditions of Contract being Clauses 1-48 as set out in the IChemE Form of Contract for Reimbursable Contracts (the 'Green Book'), 3rd edition, 2002.
   (d) The Specification.
   (e) The Schedules:
   Schedule 1: Description of the Works;
   Schedule 2: Documentation;
   Schedule 3: Responsibilities of Purchaser;
   Schedule 4: Health and Safety;
   Schedule 5: Environmental protection and waste disposal;
   Schedule 6: Quality assurance and validation;
   Schedule 7: Subcontracting;
   Schedule 8: Contractor's named personnel;
   Schedule 9: Training by Contractor;
   Schedule 10: Parts with limited working life;
   Schedule 11: Times of completion;
   Schedule 12: Liquidated damages for delay;
   Schedule 13: Pre-installation tests and procedures;
   Schedule 14: Criteria for the completion of construction;

Schedule 15: Take-over procedures;
Schedule 16: Performance tests and procedures;
Schedule 17: Performance guarantees and damages for failure;
Schedule 18: Cost elements, rates and charges;
Schedule 19: Terms of payment;
Schedule 20: Commercial records.

For the purpose of identification, the contents of the **Contract**, including the number of pages in each part, are listed in the Annex to this Agreement.

2. The **Contract** constitutes the entire agreement between the **Purchaser** and the **Contractor** with respect to the performance of the **Works** and supersedes all prior negotiations, representations or agreements relating thereto, whether written or oral, except to the extent that they are expressly incorporated in the **Contract**. No changes, alterations or modifications to the **Contract** shall be effective unless the same shall be in writing and signed by both parties.

3. The **Contractor's** liability in respect of loss of or damage to property of the **Purchaser** and his **Affiliates** in accordance with Sub-clause 30.7 of the General Conditions shall not exceed *[name of currency and amount in figures and words].*

4. The deductible for the insurance policy referred to in Sub-clause 31.2 of the General Conditions shall be the first *[name of currency and amount in figures and words]* of each and every claim.

5. The total aggregate liability of the **Contractor** to the **Purchaser** under Sub-clause 44.3 shall not exceed *[name of currency and amount in figures and words].*

6. In case of conflict between any of the documents accompanying this Agreement, the order of precedence shall be as set forth in Clause 2 of the General Conditions.

7. For the purposes of Sub-clauses 7.3, 8.2 and 28.3 of the General Conditions the date of the **Contractor's** tender shall be *[date].*

8. The date for the commencement of the **Works** shall be *[date].*

9. The **Purchaser** hereby appoints *[name]* to act as the **Project Manager** for the purposes of the **Contract**.

10. The **Contractor** hereby appoints *[name]* to act as the **Contract Manager** for the purposes of the **Contract**.

11. The bank whose base lending rate is referred to in Clause 1 of the General Conditions is *[name of bank].*

12. In accordance with Sub-clause 2.1 the law governing the Contract shall be the law of *[name of country].*

13. The language of all communications under the Contract shall be *[language].*

14. The currency of payments made under the Contract shall be *[name of currency].*

15. The exchange rate for payments made under the Contract shall be *[state here how the rate will be determined].*

16. The location of any arbitration proceedings under Clause 48 shall be *[name of place].*

17. The procedural law of any arbitration proceedings under Clause 48 shall be *[description of statute]* and the proceedings shall be conducted in *[language].*

IN WITNESS whereof the parties hereto have signed this Agreement on the date first above written

For and on behalf of *[full name of the **Purchaser**]*

Signature:

Name:

Position:

Date:

Place:

For and on behalf of *[full name of the **Contractor**]*

Signature:

Name:

Position:

Date:

Place:

Appendix A: Sample Construction Completion Certificate

Contract Title and Identification Number:

Purchaser:

Contractor:

List of parts of the Plant substantially completed, covered by this Certificate:

Criteria listed in Schedule 14 which apply to these parts of the Plant:

Programme for inspection and tests:

*We confirm that the parts of the Plant listed above are substantially complete and in such a condition that the relevant Take-Over Procedures may be safely carried out.

Signed: ....

Name: ....

Date: ....

Contract Manager

Project Manager's endorsements:

*The parts of the Plant listed above do not meet the criteria in Schedule 14 as listed above in the following significant respects:

*I require the following minor items to be completed before the issue of a Take-Over Certificate for the parts of the Plant listed above:

Signed: ....

Name: ....

Date: ....

Project Manager

*Delete as applicable

Appendix B: Sample Take-Over Certificate

Contract Title and Identification Number:

Purchaser:

Contractor:

Date of Certificate:

(Alternative A—where the Plant or any section is being taken over under Sub-clause 33.7 or 33.10)

(1) I hereby certify that the [Plant]/[section(s) of the Plant listed below], apart from any parts of the [Plant]/[section(s) listed below], [has]/[have] satisfied the requirements of the Specification and Schedule 15 of the Contract and [is]/[are] to be taken over by the Purchaser as from the taking over date stated below.

(Alternative B—where the Plant or any section is being taken over under Sub-clause 33.9)

(1) I hereby certify, with the Contractor's consent, that the [Plant]/[section(s) of the Plant listed below], apart from any parts of the [Plant]/[section(s) listed below] [is]/[are] to be taken over by the Purchaser as from the taking over date stated below.

(a)

(b)

(2) The following parts of the [Plant]/[section] are excluded from taking over for the reason(s) stated:

(a) .... because ......

(b) .... because ......

(3) The following minor items of work are still to be completed in accordance with the terms of Sub-clause 33.8 of the General Conditions:

(a) ......

(b) ......

(4) The following take-over procedures have been omitted by operation of Sub-clause [33.9]/[33.10] of the General Conditions and still remain to be carried out by the Contractor in accordance with the terms of Clause 33 of the General Conditions:

(a) ......

(b) ......

(5) The taking over date is: ....

Signed: ....

Name: ....

Project Manager

Appendix C: Sample Acceptance Certificate

Contract Title and Identification Number:

(1) I hereby certify that:

forming part/all of the (Plant)

constructed by (Contractor)

for (Purchaser)

in accordance with an Agreement and Conditions dated .... has been Accepted by the Purchaser as of the date hereof*:

(a) having passed all the required performance tests satisfactorily.

(b) having passed all the required performance tests, excepting those for which the appropriate liquidated damages have been paid by (Contractor).

(c) not having passed all its performance tests because the Contractor has been prevented by matters beyond his control from so demonstrating, but I accept the Contractor's claim that it would otherwise have passed, and in consideration of the tendering by the Contractor of a bond or guarantee in accordance with Sub-clause 35.11 of the said Conditions I therefore issue this Certificate subject to the provisions of Clause 36 of the said Conditions.

(d) following the issue of a notice of election issued to the Contractor in accordance with Sub-clause 36.6 of the said Conditions.

(2) Known Defects to be made good by the Contractor and minor items still remaining to be completed are listed in the attached 'punch list'.

Date: ....

Signed: ....

Name: ....

Project Manager

* Note: delete all but one of (1) (a), (b), (c) or (d).

Appendix D: Sample Final Certificate

Contract Title and Identification Number:

Purchaser:

Contractor:

Date of Certificate:

(1) I certify, in accordance with Sub-clause 38.1 of the General Conditions:

(a) that the Defects Liability Period has expired;

(b) that the Contractor has made good all Defects that have appeared during the Defects Liability Period in the [section of the Plant stated below]/[Plant];

(c) and that the Contractor has satisfied all its obligations (if any) under Sub-clause 38.3 of the General Conditions.

(This Final Certificate is issued in respect of the .... [section of the Plant]/[Plant])

*(2) This last Final Certificate is effective from ....

*(3) The following items are excluded from this Final Certificate under the terms of Sub-clauses 37.6 and 38.2 of the General Conditions:

(a) ....

(b) ....

Signed: ....

Name: ....

Project Manager

* If this is the last Final Certificate, (3) should be deleted. If this is not the last Final Certificate, (2) should be deleted.

Index

Abbreviations used under the heading 'other' are as follows: Agt. = Agreement; App. = Appendix; GN = Guide Note; IN = Introductory Note; Sch. = Schedule; Specn. = Specification.